Project Science

Physical Processes

Margaret Abraitis, Angela Deighan,
Brian Gallagher, Brian Smith and Michael Toner

We hope you and your class enjoy using this book. The other books in the Project Science series are:

Materials	ISBN 978 1 897675 68 7
Life Processes and Living Things	ISBN 978 1 897675 70 0
Assessing Science Key Stage 2	ISBN 978 1 897675 35 9
Science for 5–7 Year Olds	ISBN 978 1 903853 08 5
Science and Technology for the Early Years	ISBN 978 1 903853 09 2

Published by Brilliant Publications, Unit 10, Sparrow Hall Farm, Edlesborough, Dunstable, Bedfordshire LU6 2ES, UK

Sales and stock enquiries:
Tel: 01202 712910
Fax: 0845 1309300
Email: brilliant@bebc.co.uk
Website: www.brilliantpublications.co.uk

Editorial and Marketing
Tel: 01525 222292

The name Brilliant Publications and its logo are registered trademarks.

Written by Margaret Abraitis, Angela Deighan, Brian Gallagher, Brian Smith and Michael Toner
Illustrated by Virginia Gray
Cover photograph by Martyn Chillmaid

Printed in the UK

© Margaret Abraitis, Angela Deighan, Brian Gallagher, Brian Smith and Michael Toner
ISBN 978 1 897675 69 4

First published in 2000, reprinted 2007.

10 9 8 7 6 5 4 3 2

The right of Margaret Abraitis, Angela Deighan, Brian Gallagher, Brian Smith and Michael Toner to be identified as the authors of this work has been asserted by them in accordance with the Copyright, Designs and Patents Act 1988.

The following pages may be photocopied by individual teachers for class use, without permission from the publisher: 13–30, 41–67, 74–92 and 98–112. The materials may not be reproduced in any other form or for any other purpose without the prior permission of the publisher.

Contents

Introduction   5

Sheet no.   *Title*   *NC link*   *Page no.*

Electricity   7–30

Sheet no.	Title	NC link	Page no.
1	Things which use electricity	1a	13
2	Sources of electricity	1a	14
3	Building circuits	1a	15
4	Completing circuits	1a	16
5	Investigating battery-operated devices	1a	17
6	Conductors and insulators 1	1a	18
7	Conductors and insulators 2	1a	19
8	Using switches	1a	20
9	Investigating switches	1a	21
10	Controlling the brightness of bulbs	1b	22
11	Controlling the speed of a motor	1b	23
12	Controlling battery-operated devices	1b	24
13	Series circuits	1c	25
14	Parallel circuits	1c	26
15	Circuit symbols	1c	27
16	Circuit diagrams	1c	29

Forces and Motion   31–67

Sheet no.	Title	NC link	Page no.
1	Push and pull forces	2a	41
2	Force words	2a	42
3	Magnetic forces	2a	43
4	Magnetic materials	2a	44
5	Magnetic poles	2a	45
6	Magnet shapes	2a	46
7	Attraction through materials	2a	47
8	The force of gravity	2b	48
9	The force of friction	2c	49
10	Friction between different surfaces	2c	50
11	Reducing friction	2c	51
12	Air resistance 1	2c	52
13	Air resistance 2	2c	53
14	Useful air resistance	2c	54
15	Streamlining 1	2c	55
16	Streamlining 2	2c	56
17	The spinner experiment	2c	57
18	Water resistance	2c	58
19	Springs and elastic	2d	59

20	Stretched elastic	2d	60
21	The forcemeter	2e	61
22	Upthrust	2e	62
23	Direction of forces	2e	63
24	Balanced forces	2e	64
25	More balanced forces	2e	65
26	Unbalanced forces	2e	66
27	Unbalanced forces puzzle	2e	67

Light and Sound   68–92

1	A light in the dark	3a	74
2	How shadows are formed	3b	75
3	Shadow shapes	3b	76
4	Materials and light 1	3b	77
5	Materials and light 2	3b	78
6	Reflection of light	3c	79
7	Useful mirrors	3c	80
8	Reflecting light to be safe	3c	81
9	Sources of light	3d	82
10	Our eyes	3d	83
11	Making sounds	3e	84
12	Seeing sound vibrations	3e	85
13	Types of sound	3f	86
14	Sound pitch	3f	87
15	Pitch and loudness	3f	88
16	Does sound travel through air?	3g	89
17	Does sound travel through solids?	3g	90
18	Does sound travel through liquids?	3g	91
19	Stopping sound	3g	92

The Earth and Beyond   93–110

1	The shape of the Earth	4a	98
2	My mother and the pizzas	4a	99
3	The shape of the planets	4a	100
4	The size of the planets	4a	101
5	The Sun and the Earth	4b	104
6	Shadow lengths	4b	105
7	Sundial	4b	106
8	Night and day	4c	107
9	Times round the world	4c	108
10	Phases of the Moon	4d	109

Glossary   111

Welcome to what we hope is a comprehensive attempt to address the requirements of the National Curriculum (England and Wales) Key Stage 2 in a coherent way. There are three books in the series, one for each of the attainment targets:

- Life Processes and Living Things
- Materials and their Properties
- Physical Processes

Physical Processes is divided into four chapters. Each begins with background information for the teacher, summarising and explaining key concepts which students are expected to acquire.

After the background information you will find a chart listing the pupil activities, the resources needed, teaching/safety notes for each of the activities, and answers for the pupil pages. The answers are the writers' own results. In the complicated, but exciting world of science, these may be different from your results, but that does not make you wrong! Sometimes there will be small variations between the way in which you and the writers carried out the activities. This does not matter, and can make a good discussion point with pupils. The question 'why?' is a key part of science.

Pupils will need a pencil or pen for all the activities and additional sheets of paper are necessary for some. Other resources required are listed under 'Resources needed', and again at the top of each pupil page.

The 'Teaching/safety notes' give additional information and highlight areas where particular care is required. Before attempting any of the activities suggested, you should be fully conversant with the Association for Science Education's *Be Safe!* booklet and any science education safety guidelines laid down by your local authority/education department/board of governors, etc..

The bulk of the book consists of pupil sheets that may be photocopied free of charge by the purchasing institution. The sheets provide a series of exciting investigations to reinforce the concepts in the pupils' minds. The activities closely adhere to QCA's Scheme of Work for Science and provide a practical, useful way for the teacher to administer the Scheme of Work.

To make the pupil sheets easy to use we have used logos along the left margin to flag the different types of text:

 Passages the pupil should read in order to find out more about the topic

 Investigations, experiments, puzzles, and other activities

 The pupil has to write an answer down

 Activities which stretch the pupil. Many require the use of reference books and/or CD-ROMs.

The contents page indicates which section of the National Curriculum Programme of Study is addressed on each pupil sheet.

Where practicable, you should attempt the activities yourself before introducing them to your pupils. You should also ensure that the equipment you intend to use is suitable for the activity/activities and the children using it.

We must emphasize that all activities should be carried out under teacher supervision. The publishers do not accept liability for any injury or damage howsoever caused arising from activities suggested in this publication.

No scheme can give teachers all the answers. However, we hope that this scheme will enable you to use your time where it can be spent most productively – namely helping individual pupils rather than doing repetitive planning and preparing worksheets.

We hope you and your pupils will enjoy *Project Science*.

To obtain a copy of *Be Safe!* contact the Association for Science Education, College Lane, Hatfield, Hertfordshire AL10 9AA.

Electricity

Background information

Electricity is the most commonly used form of energy in the home. Electricity can be transmitted down wires from power stations to our homes. Electrical energy can be changed into other forms of energy, such as **sound** by a bell or buzzer, **light** by a bulb, **movement** by a motor, and **heat** by a heater.

Helpful words
- Electricity
- Energy
- Wire
- Bell
- Buzzer
- Bulb
- Motor
- Conductor
- Insulator
- Battery
- Solar cell
- Current
- Circuit
- Component
- Switch
- Series circuit
- Parallel circuit
- Resistance

Simple circuits

At home, the most common source of electrical energy is **electrical sockets**. These are connected through a system of cables to a power station. **Batteries** are a source of portable electrical energy and **solar cells** change light energy directly into electrical energy. Electricity flows from these sources in the form of an **electric current**.

To investigate the flow of electricity we need:
1. A source of electrical energy, such as a battery.
2. A component such as a bulb, to indicate the flow of electricity.
3. Metal wires so that we can construct a complete path of conductors from one side of the battery to the other side.

We usually think of the electrical current flowing from the negative side of the supply (−) to the positive side of the supply (+). This will only happen if there is a complete path of conductors connecting both sides of the battery. If the circuit is not complete, because of an air gap caused by an unconnected wire or component, then the electrical current will not flow.

Materials which allow an electrical current to flow through them easily are called **conductors**. All metals and some non-metals such as graphite are conductors. Materials which do not allow an electrical current to flow through them easily are called **insulators**. Air and plastics are common insulators.

Controlling electrical devices

Switches can be used to control the flow of current either by completing the circuit and allowing the current to flow or by breaking the circuit (by introducing an air gap) and stopping the flow of current. By the correct use of switches, any electrical device can be switched on or off.

When components are connected **in series** with each other, they share the electrical energy which is carried by the electric current. The more components that are connected in series, the less energy will be available for each one. In the case of bulbs, this is indicated by their brightness decreasing as more bulbs are connected in series.

Electricity

If batteries are connected in series with each other, then the amount of electrical current flowing through the circuit increases. If more than one battery is connected to a bulb, then the bulb will become brighter. However, if too many batteries are connected to a bulb, then the bulb will 'blow'. Bulbs have a maximum working voltage which is usually printed on their base. Increasing the number of batteries increases the voltage across the bulb. Two 1.5V batteries will produce 3V across a component. If the voltage across the component exceeds its working voltage, then the component could be damaged.

Connecting bulbs **in parallel** with each other connects the bulbs directly to the battery, in other words the electrical current flowing to one bulb does not have to flow through any of the others. The brightness of the bulbs is not affected by adding more bulbs to the circuit as long as they are in parallel with each other. House lights are connected in a parallel arrangement. This has the added bonus that, if one goes out or is removed, the others are unaffected.

Components can also be controlled by changing the characteristics of the wire conductor used in the circuit. Normally the wires that make up an electrical circuit are designed to have as little effect as possible on the current flowing through them. In practice, however, the electrical current loses some energy as heat as it flows through wires. The amount of difficulty experienced by the current as it flows through a wire is called **resistance**. The greater the resistance a wire has, then the smaller the size of the current flowing through it. Wires that are thin and long will have more resistance. The properties of these wires can be used to control electrical components, to regulate the brightness of a bulb or the speed of a motor, for example.

Some ideas for further work

◆ Electricity, oil and gas are almost essential for us to survive in the modern world. However, they are expensive to produce, come mostly from non-renewable resources and must be protected for future generations. Devise a plan to save some of these vital resources. For example, at home or at school, look at the sources of heating and investigate ways to stop heat being lost, such as insulation. Look also at ways to save electrical energy, such as switching off appliances when not in use. Study energy bills before and after the experiment to see if energy is being saved. (This experiment would have to be done over a long time, perhaps even a year, but it could save the school money!)

◆ Investigate alternative sources of electricity, such as wind turbines, hydroelectricity, wave and geothermal electricity, using both books and CD-ROMs.

Electricity

Activity	Answers	Resources needed	Teaching/ safety notes
Sheet 1 (page 13) Things which use electricity	**Uses electricity**: lamp, toaster, television, computer, vacuum cleaner, kettle, iron, washing machine **Does not use electricity**: desk, chair, bed, mop		
Sheet 2 (page 14) Sources of electricity	**Source**: Mains, Solar cells, Battery **Appliance**: Table lamp, Calculator, Torch Electricity reaches our homes from power stations through a network of cables called the National Grid.	Reference materials	Mains electricity is dangerous. It is safer to work with electricity from batteries.
Sheet 3 (page 15) Building circuits	• When all the parts are connected properly the bulb lights up. • When wire 1 is removed the bulb goes out. • When wire 2 is removed the bulb goes out. • When all the wires are connected back up properly, the bulb lights up. The circuit has to be complete before the bulb lights up.	Battery 2 wires Bulb Reference materials	If is **not safe** to short-circuit a battery by connecting both sides of a battery together without a component. A large current can flow which can make the battery very hot.
Sheet 4 (page 16) Completing circuits	1 This bulb does not light because both sides of the battery are not connected. 2 This bulb will not light because it has not been connected to the circuit. 3 This bulb lights because the circuit is complete. 4 This bulb will not light because both sides of the battery are not connected. 5 This bulb will light because the circuit is complete. 6 This bulb will not light because one of the wires is broken.	Battery 4 wires Bulb	Be careful not to short-circuit the batteries, especially if they are rechargeable ones.

Physical Processes

Electricity

Activity	Answers	Resources needed	Teaching/ safety notes
Sheet 5 (page 17) Investigating battery-operated devices	• The bulb could be made to produce dots and dashes by disconnecting one of the wires from the battery. • The motor can be made to turn in the opposite direction by reversing the battery. • Connecting two batteries to the bell should make it ring louder. • Buzzers normally make a noise only if the battery is connected in one direction.	Battery 2 wires Bulb Bell Motor Buzzer Reference materials	Buzzers normally have a (−) and (+) marked on them which should be matched to the (−) and (+) on the battery to make them work. Try sticking a propeller blade or a piece of card, with lines drawn on it, on to the motor axle. This will help to indicate the direction and speed that the motor is turning.
Sheet 6 (page 18) Conductors and insulators 1	Electricity passes through: aluminium can, graphite, silver coin, iron nail, brass key, copper coin.	Battery 3 wires Bulb Materials to test	All materials will eventually conduct electricity if the voltage applied across the material becomes large enough. For example, air is a good insulator but it does conduct electricity in the case of lightning. This is due to the huge voltage.
Sheet 7 (page 19) Conductors and insulators 2	**Conductor**: Aluminium can, Graphite, Silver coin, Iron nail, Brass key, Copper coin, Wool **Insulator**: Plastic ruler, Paper, Card, Wood, Cloth, Thread Materials which allow electricity to pass through them are normally metals. Graphite is a non-metal which allows electricity to pass through it easily. Non-metals normally do not allow electricity to pass through them.	Activity sheet 6	
Sheet 8 (page 20) Using switches	The paper clips act as a switch to control the bulb by introducing an air gap into the circuit to switch the bulb off. It then completes the circuit to switch the bulb on.	Battery 3 wires Bulb 2 paper clips 2 tacks Cardboard	Cardboard from packaging boxes should be thick enough to be used in this activity.

Electricity

Activity	Answers	Resources needed	Teaching/ safety notes
Activity 9 (page 21) Investigating switches	Switches are used to switch virtually all electrical appliances on and off, for example, lights, televisions, etc.	Battery 3 wires Switch Bulb Bell Motor Buzzer Reference materials	
Sheet 10 (page 22) Controlling the brightness of bulbs	• As more bulbs are added to the single battery they become dimmer. • As more batteries are added to the single bulb it gets brighter. • The brightness of the bulb can be controlled by adding more batteries or by adding more bulbs to the circuit.	3 bulbs 3 batteries 4 wires	Use a bulb designed to work with 3V, as two or three batteries could 'blow' a 1.5V bulb.
Sheet 11 (page 23) Controlling the speed of a motor	The motor in circuit 2 spins more quickly because there are two batteries in the circuit.	Motor 2 batteries 2 wires	
Sheet 12 (page 24) Controlling battery-operated devices	When a very long or thin wire is used, the bulb will be dimmer and the motor will spin more slowly.	Bulb Motor 2 batteries 2 wires 1 very long wire 1 very thin wire	Wires made out of constantan (1 metre length) should have a large enough resistance to produce a noticeable difference in the brightness and speed of the bulb and motor.
Sheet 13 (page 25) Series circuits	• Circuit 1: The bulb should be very bright. • Circuits 2 and 3: The bulbs should become dimmer and may become so dim that no light is seen at all.	Battery 4 wires 3 bulbs	
Sheet 14 (page 26) Parallel circuits	• Circuits 2 and 3: The bulbs should remain as bright as the single bulb connected to the battery in circuit 1. • Bulbs should be connected in series to make them dimmer and connected in parallel to keep them bright.	Battery 6 wires 3 bulbs	Household lights are connected in parallel. This not only keeps the voltage across them the same (230V) but has the added advantage that, if one goes out, the others can still be used.

Electricity

Activity	Answers	Resources needed	Teaching/ safety notes
Sheet 15 (pages 27 and 28) Circuit symbols	Switch = —/— Bell = (bell symbol) Motor = Ⓜ Bulb = ⊗	Resource sheet Scissors Glue	
Sheet 16 (pages 29 and 30) Circuit diagrams	1 In this circuit, electricity flows from the battery through the wires to the bulb which lights up. 2 In this circuit the bulb will light up if the swtich is on, completing the circuit. 3 In this circuit, electricity flows from the battery through the wires to the bell. It rings if the switch is on, completing the circuit. 4 In this circuit, electricity flows from the battery through the wires to the motor. It goes if the switch is on, completing the circuit.	Battery 3 wires Bulb Motor Bell	

Physical Processes
12

Things which use electricity

Electricity Sheet 1

 **Activity** Which of these things use electricity?

	Yes		No
lamp		bed	
table		vacuum cleaner	
director's chair		kettle	
toaster		iron	
television		mop	
computer		washing machine	

 **Look further** Think of other things at home which need electricity to work.

Electricity Sheet 2

Sources of electricity

Warning: electricity can kill!

 **Activity** Draw lines to match the torch, lamp and calculator below with their source of electrical energy.

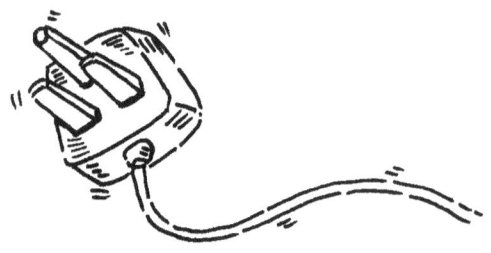

Electricity in the house

Light energy from the Sun

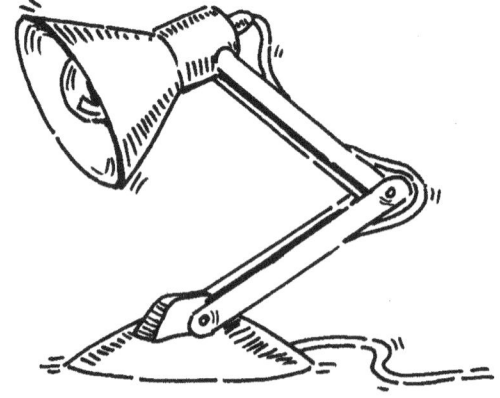

Energy from a battery

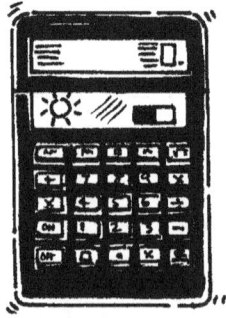

 **Look further** Find out how electricity reaches our homes.

Power station

Pylons

Building circuits

Electricity Sheet 3

You need: a battery, 2 wires and a bulb.

Activity

Look at the drawing below. It shows some electrical parts connected together.

◆ Collect a battery, two wires and a bulb.

◆ Set up the parts as shown in the diagram.

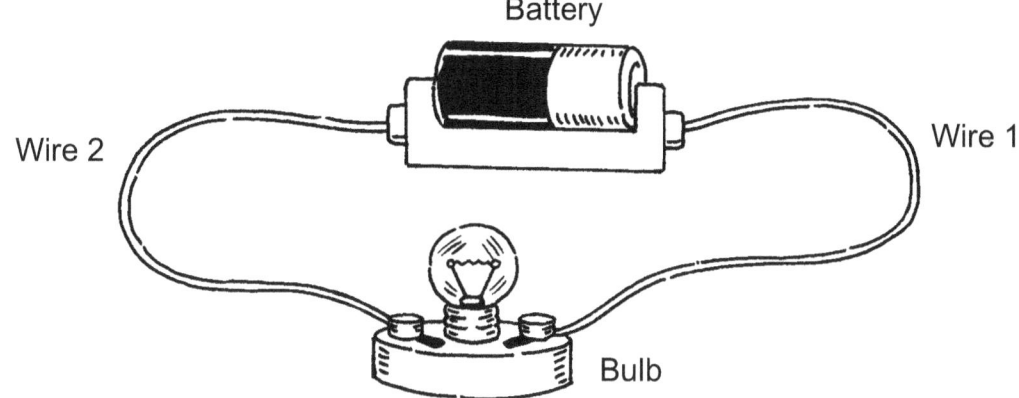

Answer What happens when you have connected all the parts?

What happens if you take away wire 1?

Reconnect wire 1. Take away wire 2. What happens?

Reconnect wire 2. Take out the battery. What happens?

Look further Explain why the bulb goes on and off.

Completing circuits

Electricity Sheet 4

You need: a battery, 4 wires and a bulb.

Activity In which of these circuits will the bulb light? Build the circuits to check your answers. Cross out *will* or *will not* and give a reason for each answer.

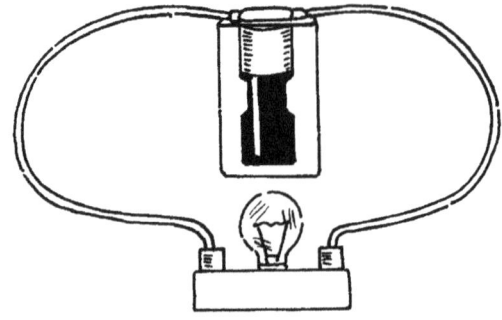

1 This bulb will/will not light because...

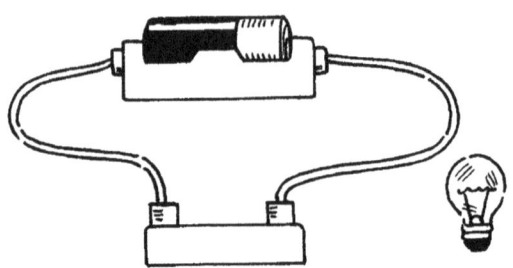

2 This bulb will/will not light because...

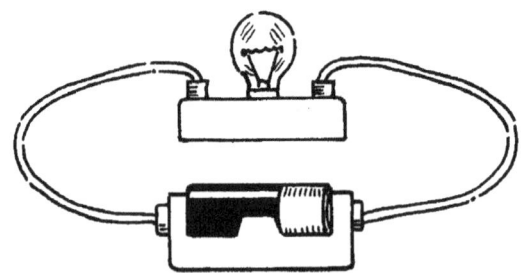

3 This bulb will/will not light because...

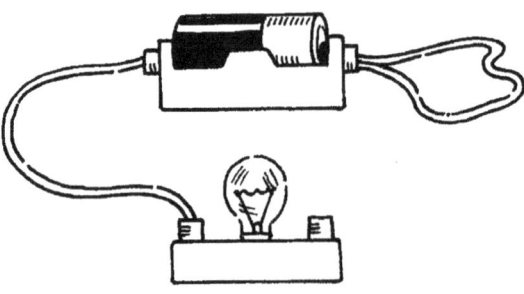

4 This bulb will/will not light because...

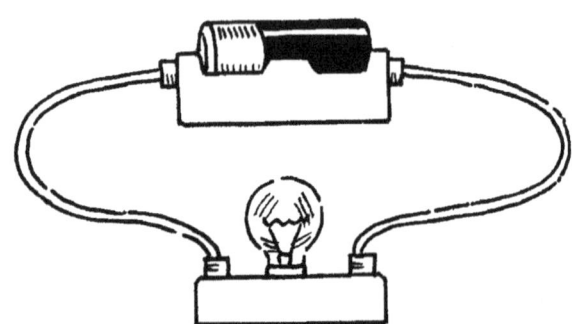

5 This bulb will/will not light because...

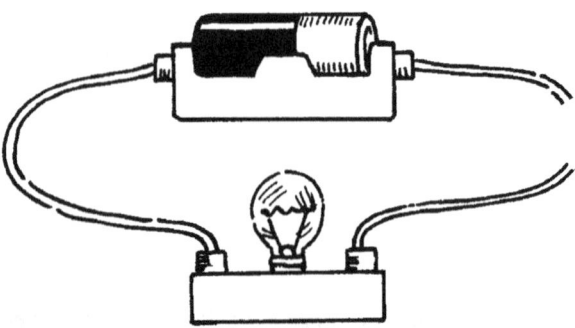

6 This bulb will/will not light because...

Electricity Sheet 5

Investigating battery-operated devices

You need: a battery, 2 wires, a bulb, a bell, a motor and a buzzer.

Activity

Use your knowledge of circuits to build and investigate uses for these circuits.

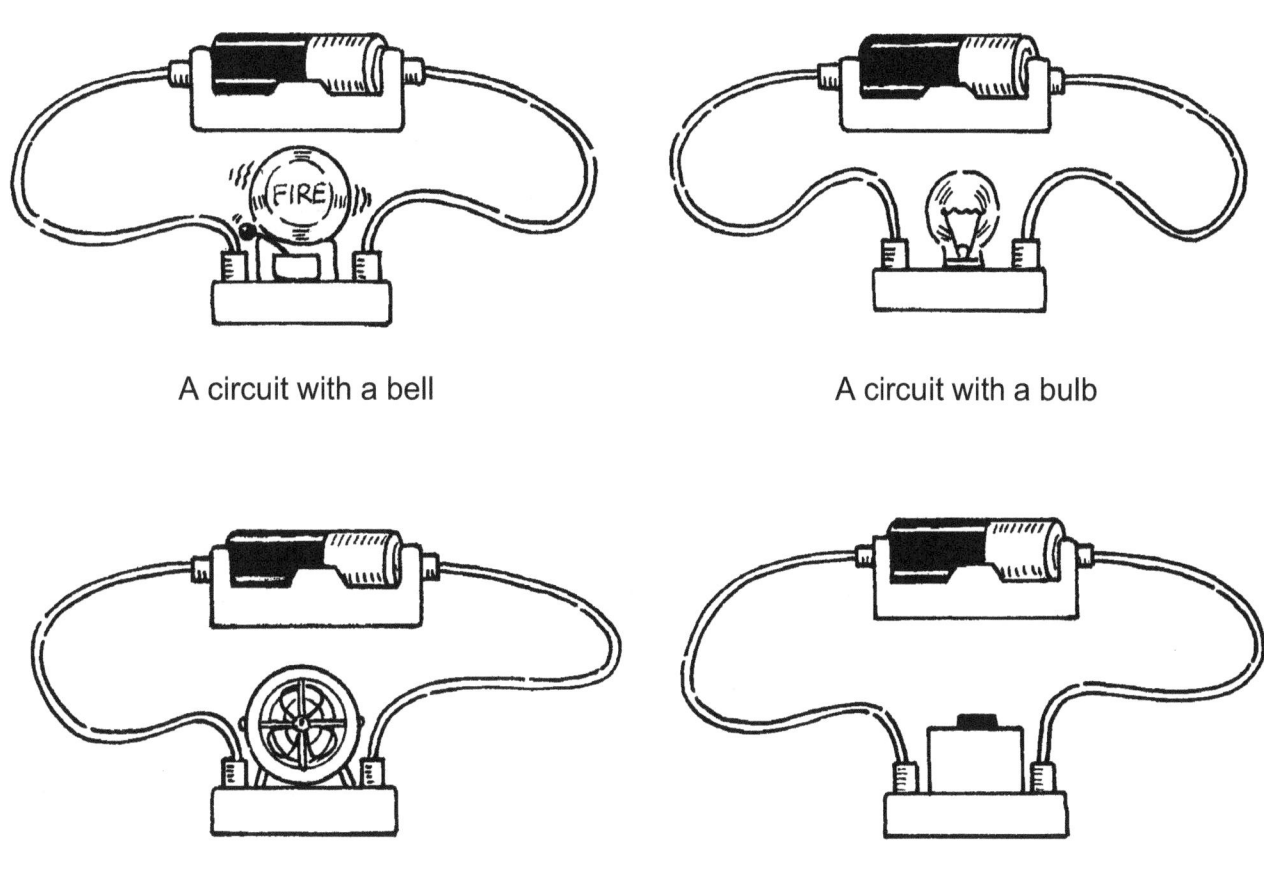

A circuit with a bell

A circuit with a bulb

A circuit with a motor

A circuit with a buzzer

Look further

Investigate how you could:
- use a light to communicate in code
- make a motor turn in the opposite direction
- make a bell ring louder.

Does a buzzer make a noise if the battery is reversed?

© M Abraitis, A Deighan, B Gallagher, B Smith & M Toner
This page may be photocopied for use by the purchasing institution only.

Electricity Sheet 6 — Conductors and insulators 1

You need: a battery, 3 wires, a bulb and some materials to test.

Read

Materials which allow electricity to pass through them are called **conductors**. Materials which do not allow electricity to pass through them are called **insulators**.

Activity

Some materials allow electricity to pass through them, others do not.

Set up the following circuit.

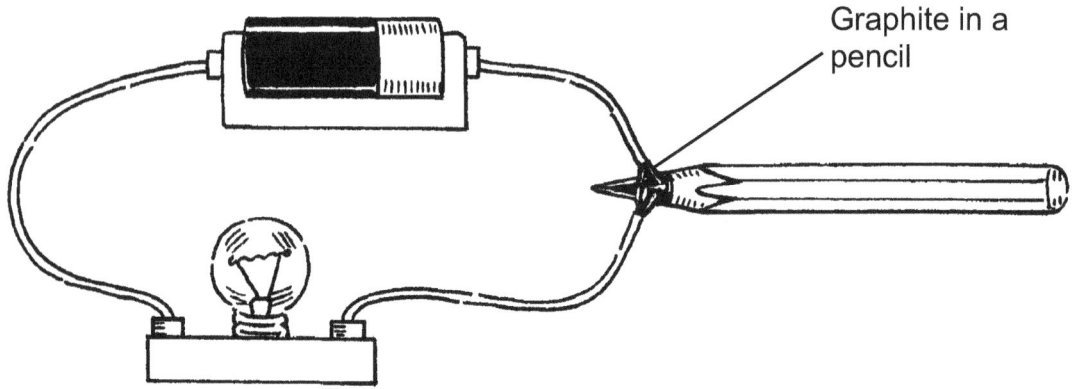

Graphite in a pencil

◆ Collect all the materials you are going to test. (Use the last two boxes for materials of your own choice.)

◆ Place each item in turn in the gap between the wires and watch what happens.

◆ Tick the materials which allow electricity to pass through them.

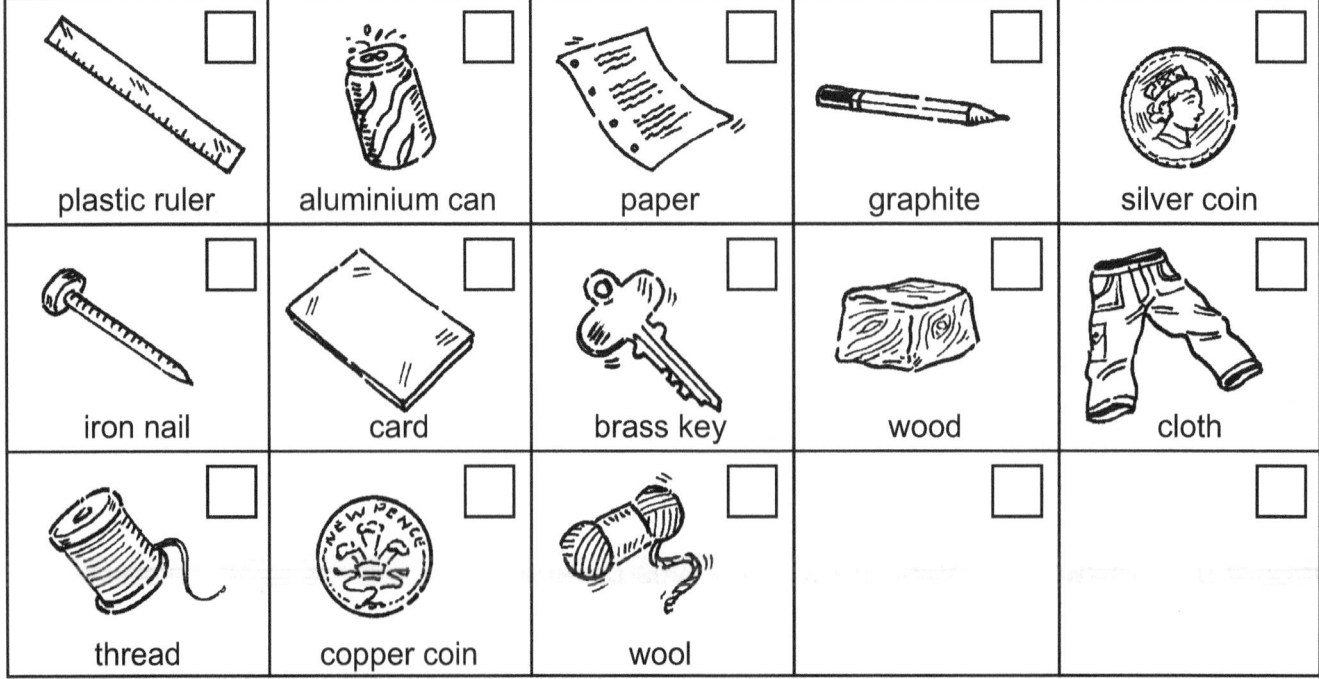

plastic ruler	aluminium can	paper	graphite	silver coin
iron nail	card	brass key	wood	cloth
thread	copper coin	wool		

Physical Processes
18

© M Abraitis, A Deighan, B Gallagher, B Smith & M Toner
This page may be photocopied for use by the purchasing institution only.

Conductors and insulators 2

Electricity Sheet 7

You need: Sheet 6 (Conductors and insulators 1).

Look at the results from the conduction experiment. Write the materials in the correct column below.

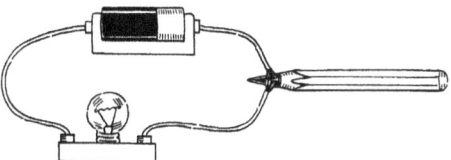

Conductor	Insulator

What do you notice about the materials which do allow electricity to pass through them?

What do you notice about the materials which do not allow electricity to pass through them?

Electricity Sheet 8: Using switches

You need: a battery, 3 wires, a bulb, 2 paper clips, 2 tacks and some cardboard.

Read

You will have already made a circuit for lighting a bulb. The bulb remains lit as long as the components (the things which are used) are connected in this circuit.

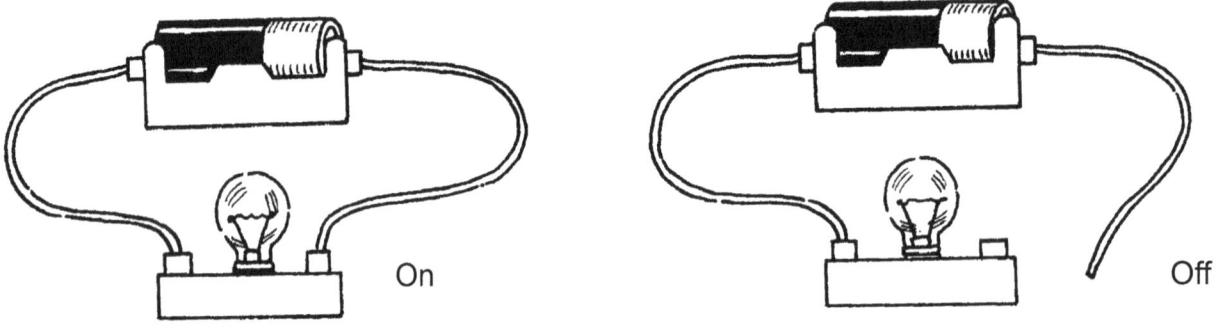

On Off

We can use a switch to turn the bulb on and off instead of attaching and detaching one of the wires.

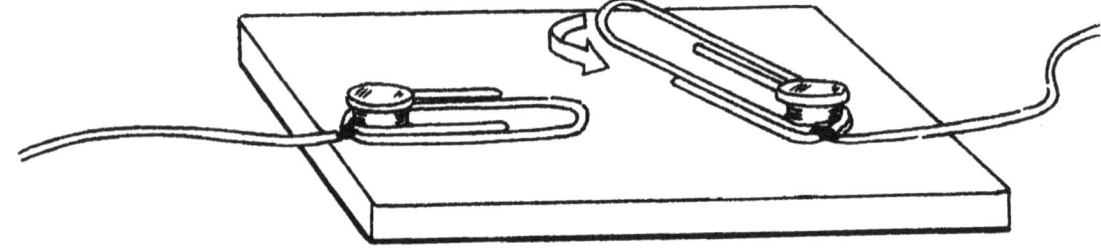

Activity

- Use paper clips and tacks to make a switch for controlling the bulb.
- Draw a circuit using a switch which will allow you to switch the bulb on and off.

Physical Processes 20

© M Abraitis, A Deighan, B Gallagher, B Smith & M Toner
This page may be photocopied for use by the purchasing institution only.

Electricity Sheet 9

Investigating switches

> You need: a battery, 3 wires, a switch, a bulb, a bell, a motor and a buzzer.

Use your knowledge of circuits to build and investigate the circuits shown below.

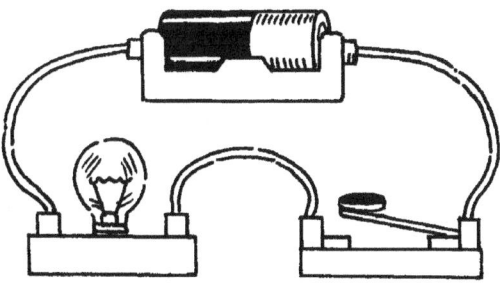

A circuit with a bulb

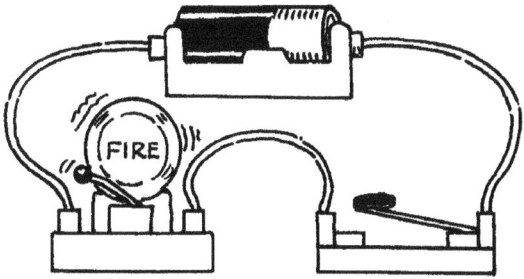

A circuit with a bell

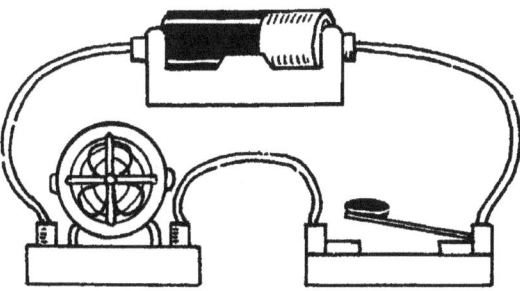

A circuit with a motor

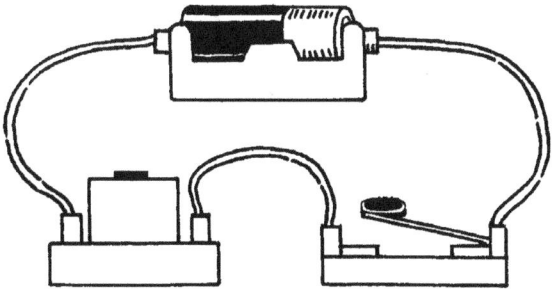

A circuit with a buzzer

Think of other circuits which use switches in a useful way. Draw one below.

Electricity
Sheet 10

Controlling the brightness of bulbs

You need: 3 bulbs, 3 batteries and 4 wires.

 **Activity** Connect up each of the circuits below and observe the brightness of the bulbs in each circuit.

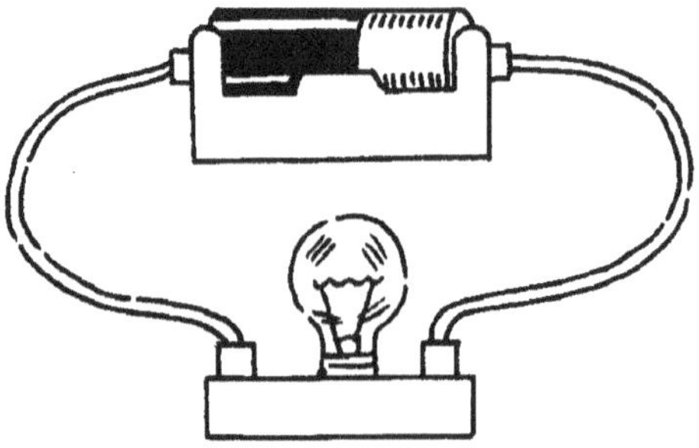

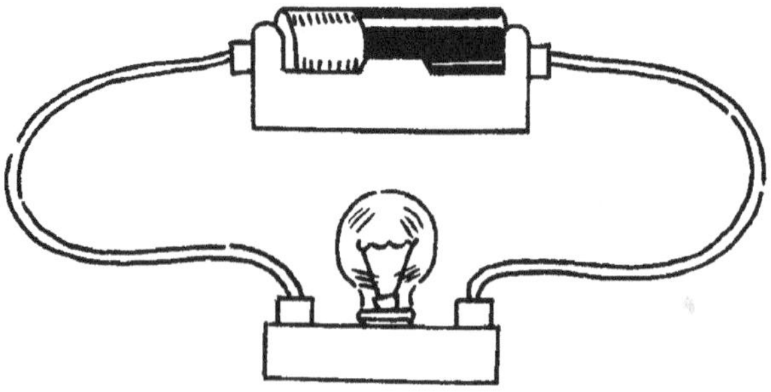

	Change the direction of the battery
Add more bulbs	Add more batteries
Add even more bulbs	Add even more batteries

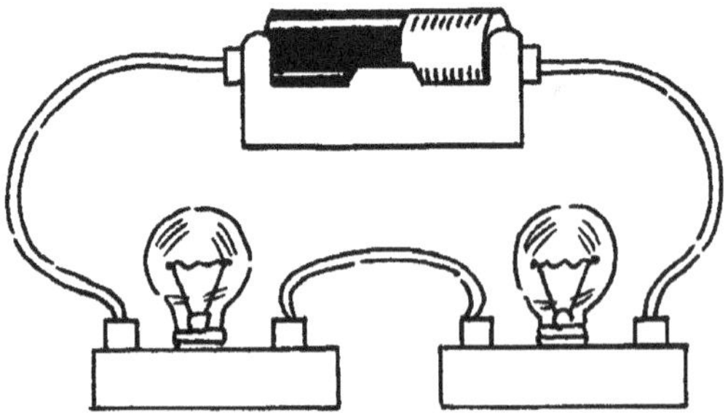

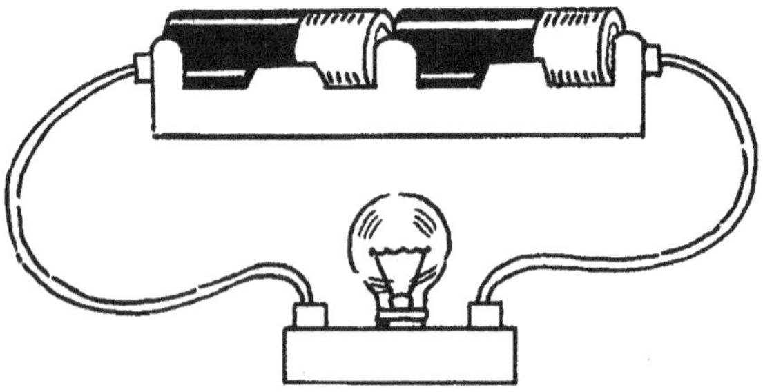

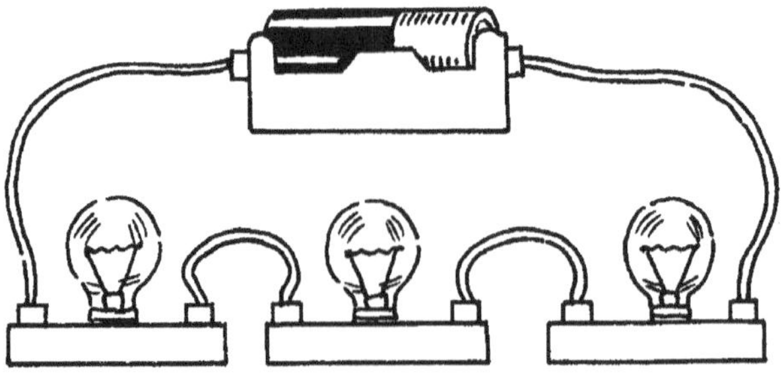

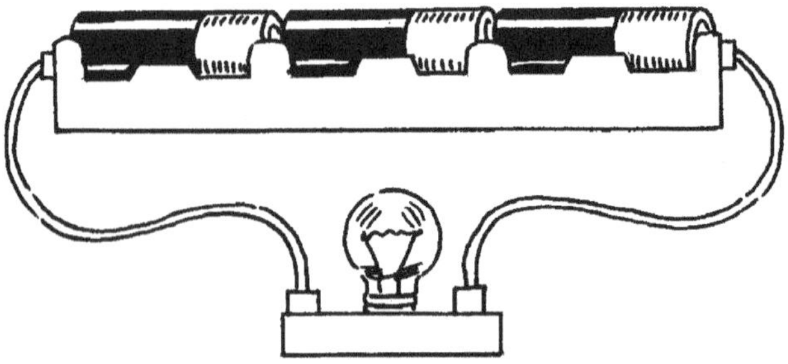

 **Answer** How can you control the brightness of the bulbs?

Physical Processes

Electricity Sheet 11

Controlling the speed of a motor

You need: a motor, 2 batteries and 2 wires.

 Make these circuits.

Does the motor spin more quickly in circuit 1 or circuit 2?

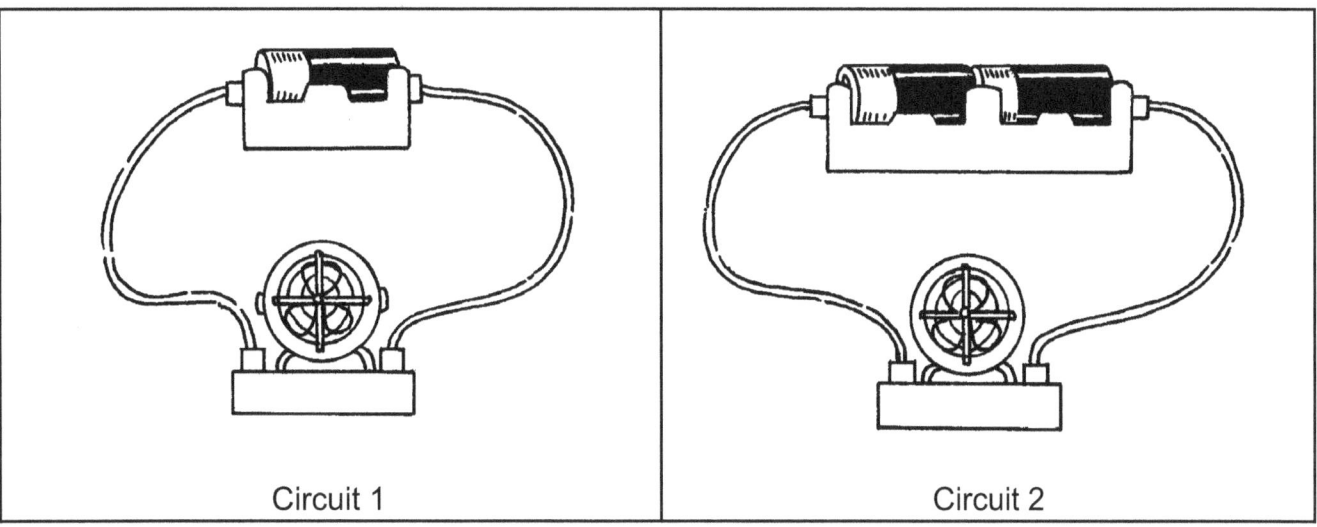

Circuit 1   Circuit 2

 The motor in circuit ____ spins more quickly because...

 Draw the circuit with the quicker motor.

Electricity Sheet 12 — Controlling battery-operated devices

You need: a bulb, a motor, 2 batteries, 2 wires, a very long wire and a very thin wire.

Activity: Investigate what happens to the speed of the motor and the brightness of the bulb when you change the type of wires.

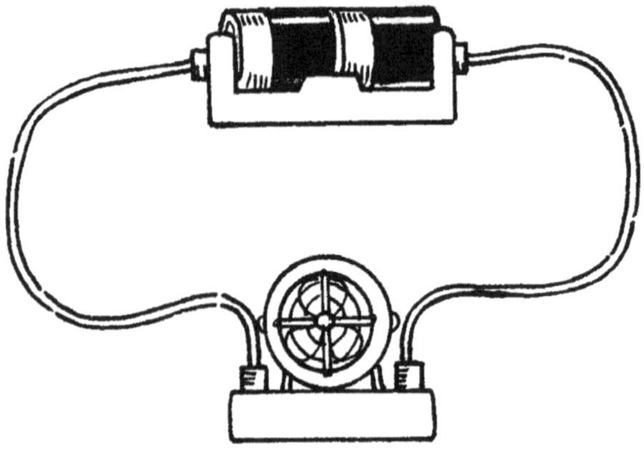

A standard wire	A standard wire
A very long wire	A very long wire
A very thin wire	A very thin wire

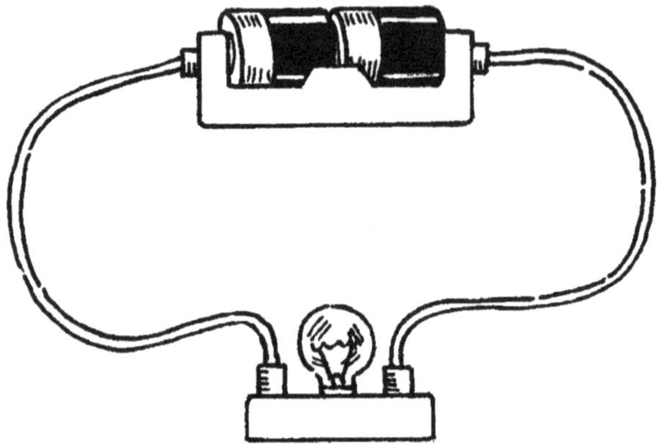

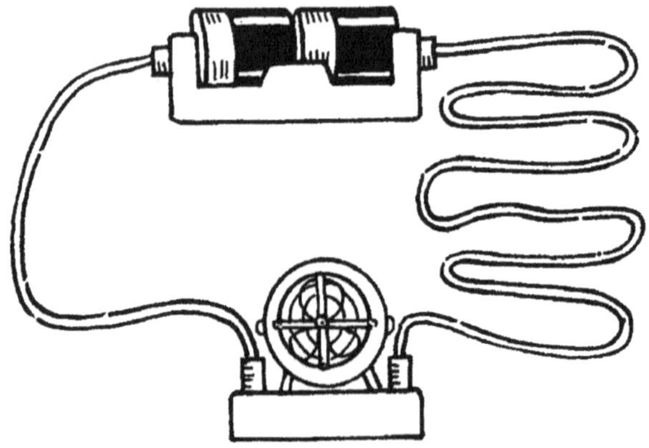

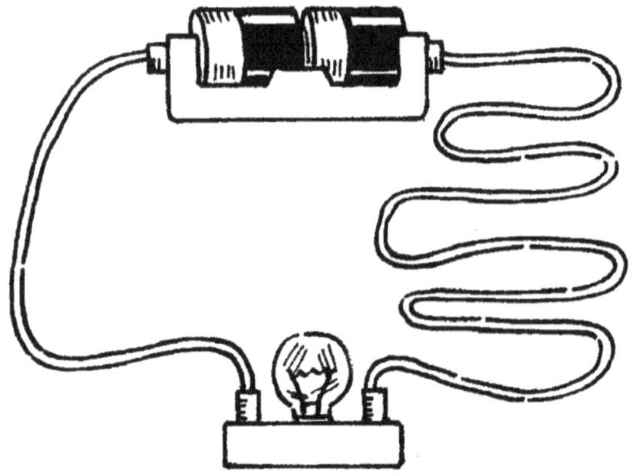

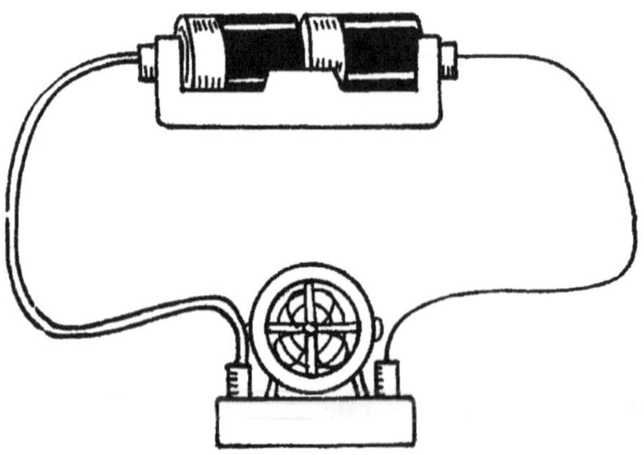

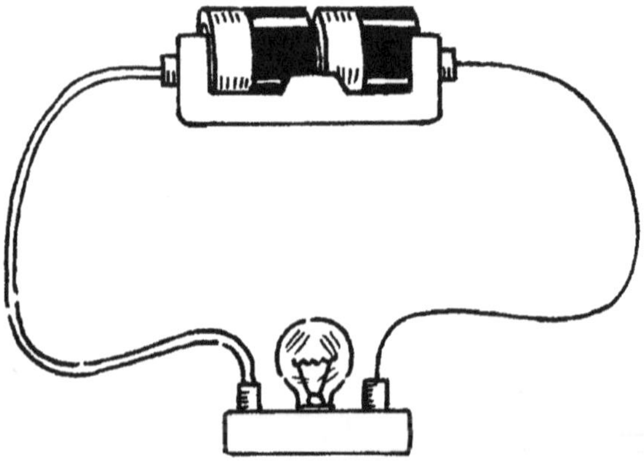

Physical Processes
24

© M Abraitis, A Deighan, B Gallagher, B Smith & M Toner
This page may be photocopied for use by the purchasing institution only.

Series circuits

Electricity Sheet 13

You need: a battery, 4 wires and 3 bulbs.

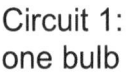

◆ Use the components listed to set up each of the circuits shown below.

◆ Decide if the bulbs are **very bright**, **bright** or **dim**.

Circuit 1:
one bulb

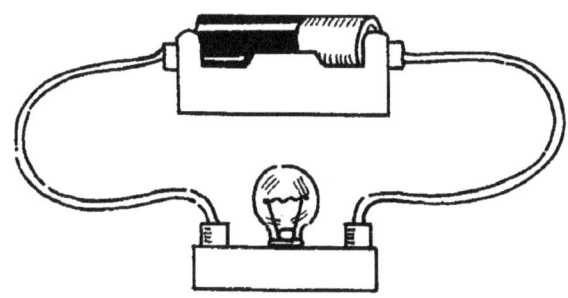

This bulb is _____.

Circuit 2:
two bulbs in series with each other

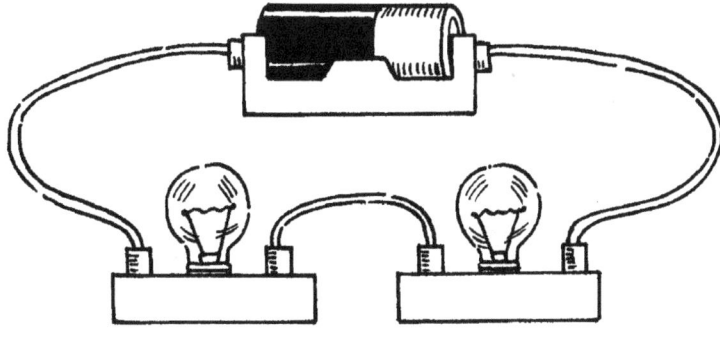

These bulbs are _____.

Circuit 3:
three bulbs in series with each other

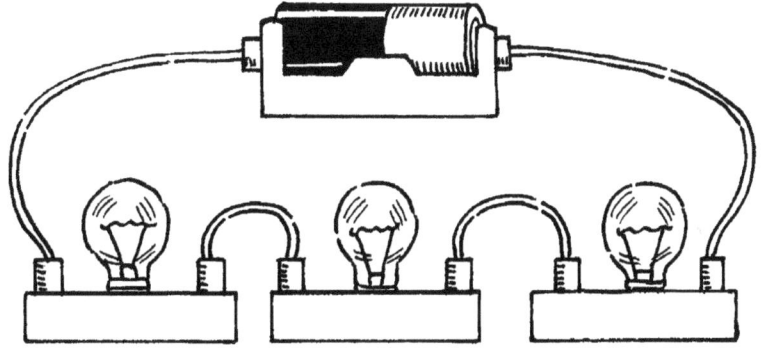

These bulbs are _____.

| Electricity Sheet 14 | # Parallel circuits |

You need: a battery, 6 wires and 3 bulbs.

- Use the components listed to set up each of the circuits shown below.
- Decide if the bulbs are **very bright**, **bright** or **dim**.

Circuit 1: one bulb

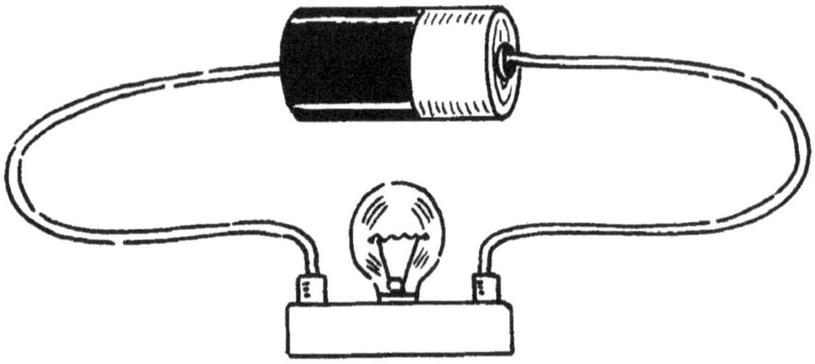

This bulb is _____.

Circuit 2: two bulbs in parallel

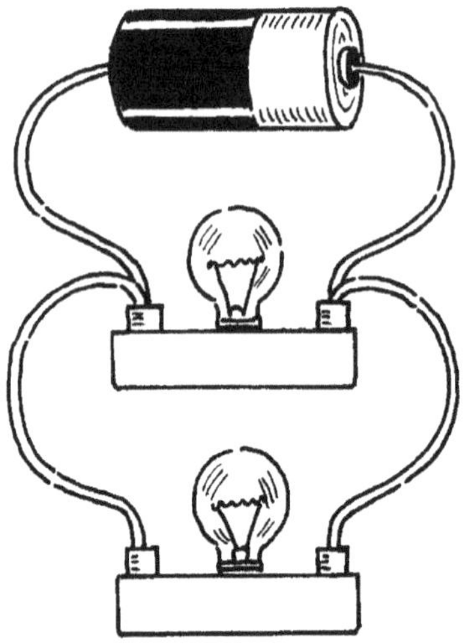

These bulbs are _____.

Circuit 3: three bulbs in parallel

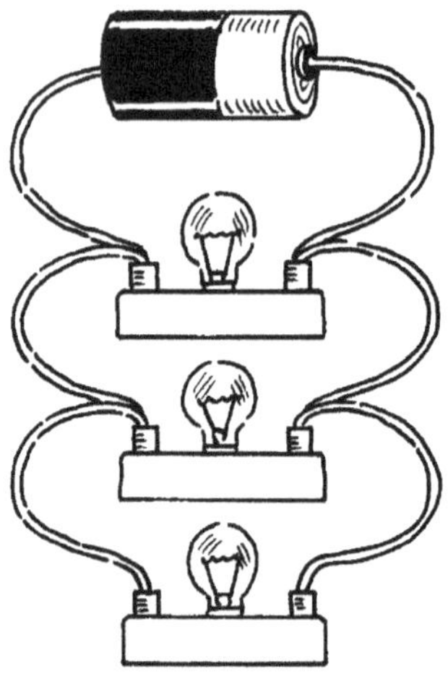

These bulbs are _____.

From your work on series and parallel circuits, describe how you would connect bulbs to a battery to make them dim and to keep them bright.

Circuit symbols

Electricity Sheet 15

You need: Resource sheet (Circuit symbols), scissors and glue.

Read

Lots of different people use electric circuits as part of their jobs, for example electricians, electrical engineers and scientists. It would take far too long to design circuits if you always had to draw the real electrical parts. To make the process much quicker, we use circuit diagrams, with circuit symbols to represent the real electrical parts.

Activity

◆ Look at the circuit symbols on the resource sheet.

◆ Cut and paste the correct symbols from the sheet to complete the table below. The first one is done for you.

Component drawing		Symbol
Battery	(battery image)	⊣⊢
Switch	(switch image)	
Bell	(fire bell image)	
Motor	(motor image)	
Bulb	(bulb image)	

Circuit symbols

Electricity
Sheet 15 – Resource sheet

You will not meet many of the component symbols below until you are older. The ones you may meet in your present school are shaded.

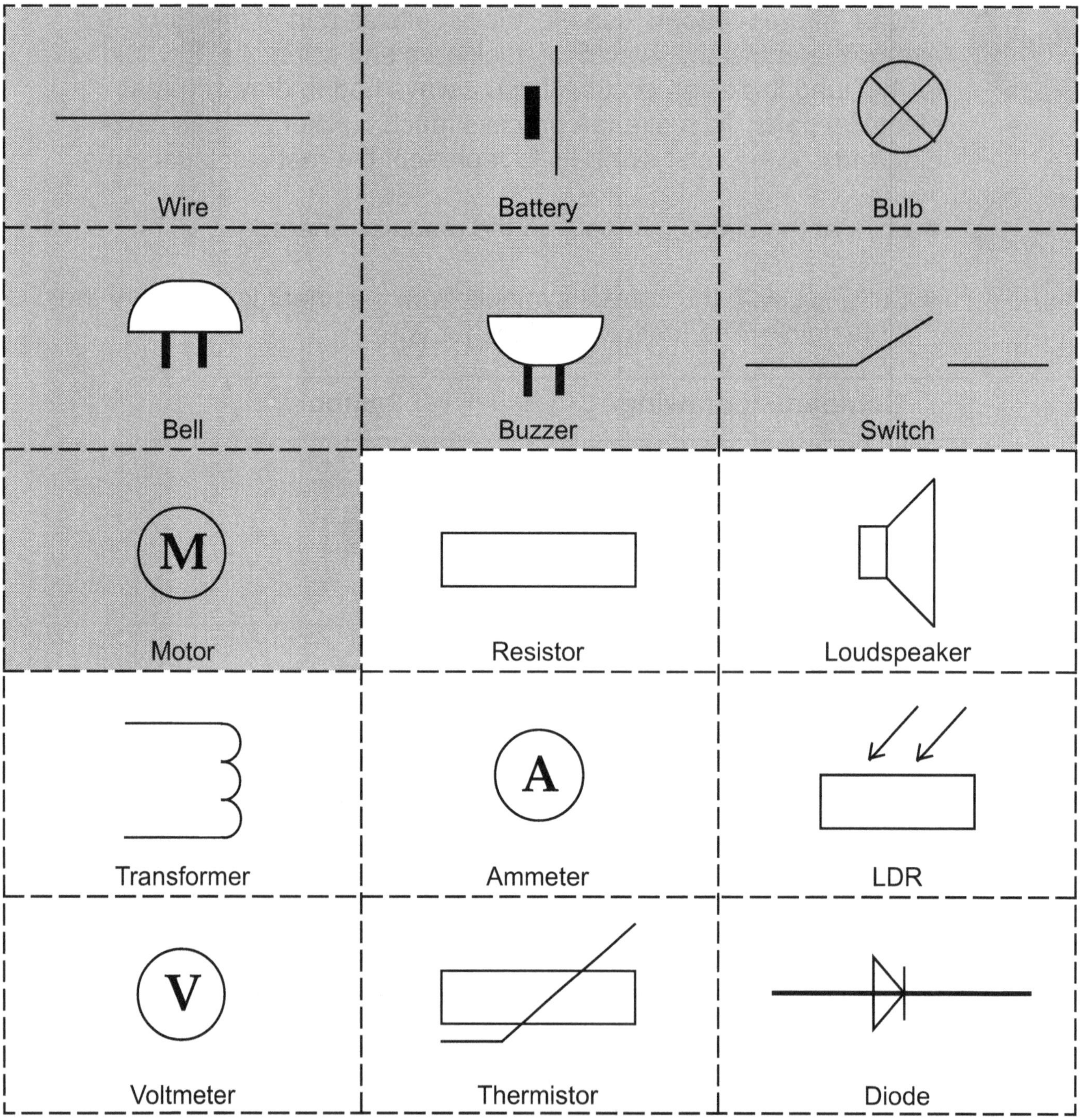

Wire	Battery	Bulb
Bell	Buzzer	Switch
Motor	Resistor	Loudspeaker
Transformer	Ammeter	LDR
Voltmeter	Thermistor	Diode

 Find out what some of the other components do.

Circuit diagrams

Electricity Sheet 16

You need: a battery, 3 wires, a bulb, a motor and a bell.

◆ For each drawing, draw the corresponding circuit diagram using the circuit symbols.

◆ Set up the circuit and explain how the circuit works. The first one is done for you.

Circuit 1

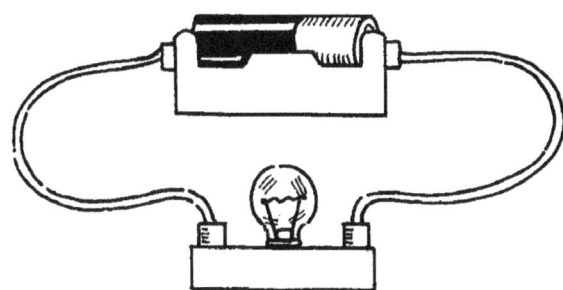

Drawing of electrical parts

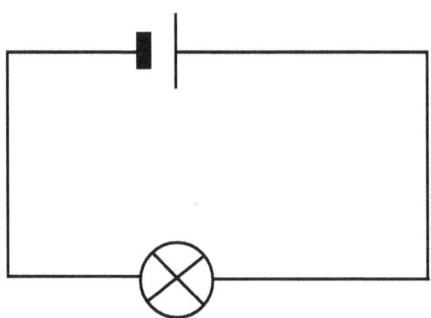

Circuit diagram

In this circuit, electricity flows from the _____ through the _____ to the _____ which lights up.

_____

Circuit 2

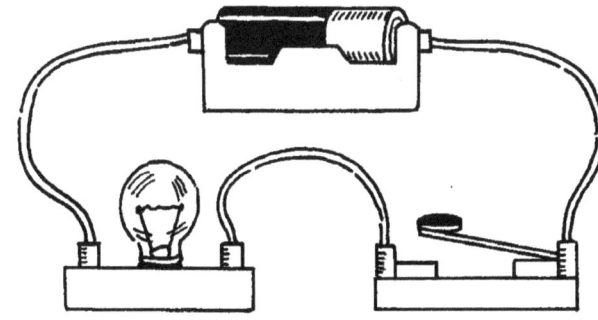

Drawing of electrical parts

Circuit diagram

In this circuit, ...

Circuit diagrams

Electricity
Sheet 16 – Continued

Circuit 3

Drawing of electrical parts

Circuit diagram

In this circuit, ...

Circuit 4

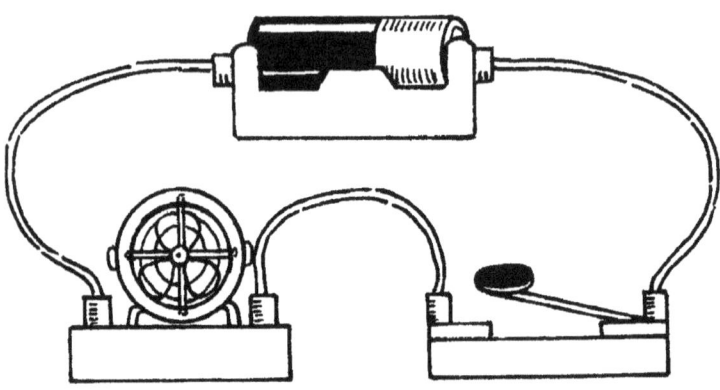

Drawing of electrical parts

Circuit diagram

In this circuit, ...

Forces and Motion

Background information

Magnetism

There are a few naturally occurring materials which are attracted to magnets. They are iron, nickel and cobalt, which are all metals. Steel is made from iron and so objects made from or containing steel will be attracted to magnets. Most of the metal objects in a room will consist of steel, including the legs of desks and chairs, scissors, sharpener blades and radiators. Coins do not normally contain iron, nickel or cobalt, so they will not be attracted to magnets. Some new coins have a steel middle which is surrounded by a coating of copper.

Magnets produce an invisible field around them called a **magnetic field** which permanently stretches out from the magnet. The strength of this magnetic field diminishes very quickly the further it stretches from the magnet. It is the magnetic field which produces the force that causes certain materials to be attracted to the magnet. The magnetic field stretches around the magnet from end to end. These ends are called **poles**. It is at the poles that the magnetic field is the strongest. The Earth produces a magnetic field that is very similar to the magnetic field of a bar magnet.

The magnetic fields of different magnets cause a force of attraction between unlike poles (North and South, South and North) and a force of repulsion between like poles (North and North, South and South). Only magnets can repel other magnets. Magnetic materials, such as iron, can only be attracted to magnets and not be repelled by them.

Helpful words	
Attraction	Friction
Repulsion	Air resistance
Pole	Streamlining
Gravity	Upthrust
Forcemeter	Compressed

Equal forces

Often there can be a force acting on an object which has no noticeable effect on the object. For example, **gravity** is constantly pulling us down. We stay where we are because the ground we are standing on holds us up by exerting an upward force on us. In most cases the upwards force is exactly equal to the force of gravity but opposite in direction. In this way the forces cancel each other out and we call these balanced forces. When the forces on a stationary object are balanced, then the object will stay stationary. Our weight is the force of the Earth's gravity pulling us downwards.

Friction

The force of **friction** occurs when two surfaces rub against each other. Friction is a force which tends to oppose any moving object's motion by slowing it down. For example, if you push a coin across the table it will soon come to a halt. This is because of the friction between the coin and the table's surface. Friction can be reduced by the use of wheels or rollers. These reduce the area of the two materials rubbing together, thus enabling the surfaces to peel away from each other,

Forces and Motion

rather than rub against each other. This can be seen by rolling the coin across the table rather than pushing it.

Air resistance

Air resistance is caused by the surface of a moving object moving through the air. As the air rubs against the moving surface, a force is produced which pushes against it. Air resistance is like friction in that it is caused by rubbing and, therefore, opposes motion. The amount of air resistance pushing against a moving object depends mostly on the speed of the moving object and its surface area. The faster the object, then the greater will be the air resistance pushing and/or pulling against it. A paraglider can capitalize on air resistance to make a jump last longer by using a parachute with a large surface area which can change its shape to capture as much air underneath as possible. By increasing the amount of air resistance, the parachutist can slow down to a low speed and prolong the jump.

A sail has a large surface area which is designed to trap the moving air behind it. This trapped air pushes the sail forward. Many sails are designed to be angled so that the surface area facing into the air can be changed to make it possible to control the speed and direction of the boat.

Streamlining is a term used to describe the process of reducing the amount of air resistance acting on an object. Streamlining can be achieved by minimizing the surface area facing into the air, using smooth, polished surfaces, and by making the surfaces curved so that air flows around them smoothly to reduce resistance.

Opposing forces

Forces always occur in pairs. For example, if you kick a heavy object forward then you feel the object exert a force on your foot. A car moves forward because the tyres push the road surface backwards and, in return, the road surface pushes the car forward. When an elastic band is stretched, it exerts a force in the opposite direction but this is equal in size to the force causing the stretch. When a spring is compressed it also exerts a force in the opposite direction but equal in size to the force causing the compression. A propeller-driven boat works by pushing water back and, in return, the water pushes the propeller (and boat) forward.

Forces always act in one direction. For example, the force of gravity always acts downwards and upthrust from liquids like water always acts upwards. All the forces can be described as either pushes or pulls. For example, we push children's swings to force them forward and we pull doors closed. Both these phrases describe

Forces and Motion

forces. We often use alternatives to 'push' and 'pull' to describe forces. Some of these are twist, tear, drag, squash and squeeze.

Springs are good for measuring the size of pulling forces because they stretch when pulled. The length of the stretch is proportional to the size of the pulling force, that is, if the pulling force is doubled the stretch of the spring will double. A good spring will also return to its normal length when the force is removed.

Some ideas for further work

- Design a streamlined vehicle using knowledge of resistance and battery-powered circuits. Balsa wood, card, glue and wheels could be used to construct the vehicle. The vehicle could be designed to travel on land or water. Modern designs such as the hovercraft could be used.

- Plan and test a way of reducing friction by using rollers to move a heavy weight.

- Find out about lodestones and how they were used to make early compasses. Where were the first compasses used?

- Collect pictures of naturally streamlined animals such as fish and birds.

- Who was Sir Isaac Newton and what was his connection with forces?

Forces and Motion

Activity	Answers	Resources needed	Teaching/ safety notes
Sheet 1 (page 41) Push and pull forces (revision)	1 push 2 push 3 push 4 pull 5 push 6 push 7 pull 8 both 9 pull		
Sheet 2 (page 42) Force words (revision)	Twist, tear, drag, squash, squeeze. Extra force words: tug, yank.		
Sheet 3 (page 43) Magnetic forces	When a North and South pole are brought together, we feel them **pull** each other **together**. When two North or two South poles are brought together, we feel them **push** each other **apart**. The push and pull demonstrate the force involved.	2 bar magnets 2 horse-shoe magnets	Like poles repel and unlike poles attract.
Sheet 4 (page 44) Magnetic materials	One end of the magnet and the paper clips will be attracted to the magnet. If the other metal objects contain iron (such as steel) then they will also be attracted to the magnet. Only the metals should be attracted to magnets.	Magnet Materials to test	
Sheet 5 (page 45) Magnetic poles	More paper clips should be picked up at points 1 and 5, with the number of paper clips picked up at these points being equal. These ends are called poles and the magnet is strongest at these points. The magnet becomes weaker as you move closer to the middle. No paper clips should be picked up at the centre of the magnet (3).	Bar magnet Paper clips	

Physical Processes

Forces and Motion

Activity	Answers	Resources needed	Teaching/ safety notes
Sheet 6 (page 46) Magnet shapes	• The magnet which can attract the paper clip from the furthest distance is the strongest. • The magnet which can hold the most paper clips at its poles is the strongest.	Magnets (different shapes) Ruler Paper clips	
Sheet 7 (page 47) Attraction through materials	A strong magnet will attract a paper clip through a thick material such as wood. The magnet should attract the paper clip through all these materials although its attraction will be reduced if the magnetic field has to pass through another magnetic material.	Strong magnet Paper clip Materials to test	
Sheet 8 (page 48) The force of gravity	The values below are only a guide. The values will depend on the type of elastic band and the mass of the object. **Object** **Elastic band length** tin of soup 12 cm adult shoe 16 cm book 8 cm jacket 14 cm pint of water 22 cm	Strong plastic bag Large elastic band String Ruler Objects to test (different sizes and weights)	The heavier the object then the longer the elastic band will stretch. This is a good opportunity to establish the relationship between the stretch of an elastic band or spring and the size of the forces acting on it.

Physical Processes

Forces and Motion

Activity	Answers	Resources needed	Teaching/ safety notes
Sheet 9 (page 49) The force of friction	• Friction between the wheels and ground slow the skate boarder down. Friction here is not useful. • Friction between the climber's shoes and the stone help hold her onto the cliff. Friction is useful here. • Friction between the tyres and the road help prevent the bike from skidding. Friction is useful here. • Friction between the moving parts of the engine cause them to wear down and overheat. Friction here is not useful.		Oil is used to reduce the force of friction between moving machinery. The oil forms a smooth liquid layer between the moving parts of the machinery to prevent them rubbing against each other.
Sheet 10 (page 50) Friction between different surfaces	The rubber will produce the greatest force of friction.	Piece of card Rubber Plastic pen Cardboard box Soft toy	An alternative to this activity would be to pull a shoe with a forcemeter over different surfaces, such as carpet, tarmac, wood, etc. It may be necessary to put weights into the shoe to keep it steady.
Sheet 11 (page 51) Reducing friction	• The car begins to slow down as soon as the pushing force is removed. This demonstrates that friction is a force which slows moving objects down. • The car or truck with wheels which are not free to turn will be the hardest to pull because the force of friction acting beween the bottom of the truck and the floor will be greater.	Toy car Toy truck Elastic band/forcemeter	Friction is the force which slows the vehicle down. Friction is created by the wheels rubbing against the floor surface. As a guide, the force needed to pull a truck with wheels is about 2 newtons; the force needed to pull a truck without wheels is about 5 newtons.

Forces and Motion

Activity	Answers	Resources needed	Teaching/ safety notes
Sheet 12 (page 52) Air resistance 1	• The feather will fall due to the pull of gravity. It will, however, fall slowly due to air resistance acting against its motion. The feather may move from side to side as it falls due to variations in air resistance. • As the card is wafted slowly upwards and downwards, resistance should be felt to its motion. This force of resistance may cause the end of the cardboard to bend in the opposite direction of its movement. The size of this resistance depends on the speed of the cardboard.	Feather Stiff card	
Sheet 13 (page 53) Air resistance 2	The two pieces of paper have the same mass but have different surface areas. The paper with the larger surface area will normally take longer to fall the same height as the crumpled-up piece of paper, which has the smaller surface area. This is because the crumpled-up piece of paper experiences less air resistance as it falls through the air. Air becomes trapped underneath the flat piece of paper and pushes against its motion. Air can flow easily around a tightly crumpled-up piece of paper, having little effect on its motion.	4 pieces of paper	The greater the height from which the pieces of paper are dropped, the more apparent the time difference should be.

Physical Processes

Forces and Motion

Activity	Answers	Resources needed	Teaching/ safety notes
Sheet 14 (page 54) Useful air resistance	• The parachutist is using the large surface area of the parachute to trap air. • The trapped air pushes against the parachute's motion and slows it down. This is air resistance.	Paper Scissors Thread Plasticine Stop clock/watch	
Sheet 15 (page 55) Streamlining 1	The streamlined animals are the greyhound, gazelle, horse and the hawk. The animals which are not streamlined are the elephant, pig, dog, sheep, turkey, bull, hippopotamus and the cow.	Scissors Glue	Many animals can change their shape to either reduce or increase air resistance, for example swans landing and hawks diving.
Sheet 16 (page 56) Streamlining 2	The vehicles are streamlined because they have round edges and smooth surfaces. The front of each vehicle is designed to allow air to flow around the vehicle.	Reference materials	
Sheet 17 (page 57) The spinner experiment	If the spinner does not spin when it falls then try clipping half a centimetre off the length of its wings.	Scissors Paper Paper clips Metre stick Timer	The winged fruit of sycamore trees are good examples of a spinner in nature.
Sheet 18 (page 58) Water resistance	The long thin shape (in the middle) will sink the fastest. The wide flat shape will sink the slowest.	Plasticine Timer String Wallpaper paste Cylinder	The children could look at photographs of submarines or marine animals such as dolphins, seals and sharks.
Sheet 19 (page 59) Springs and elastic	1 The stretched bow string will push the arrow forwards. 2 The stretched elastic will push the stone forwards. 3 The compressed spring in the pogo stick will push the girl upwards. 4 The stretched spring board will push the diver upwards. 5 The stretched trampoline skin will push the man upwards. 6 The stretched bungee rope will pull the girl upwards.		

Forces and Motion

Activity	Answers	Resources needed	Teaching/ safety notes
Sheet 20 (page 60) Stretched elastic	The further the elastic is stretched, the more energy it will store. This energy is transferred to the car when released.	Toy car Ruler Board with 2 screws Large elastic band	The values obtained in the experiment will depend on the elastic, the mass of the car and the surface of the board.
Sheet 21 (page 61) The forcemeter	The forcemeter has a spring which stretches when it is pulled. The greater the pulling force, the more the spring will stretch.	Forcemeter String Materials to test	As a guide, opening a drawer takes a force of about 15 newtons, pulling books 3 newtons, loosening a shoe lace 24 newtons, and lifting a jacket 9 newtons.
Sheet 22 (page 62) Upthrust	**Mass Weight Weight** **in air in water** 100 g 1.0 N 0.85 N 200 g 2.0 N 1.65 N 300 g 3.0 N 2.50 N 400 g 4.0 N 3.30 N 500 g 5.0 N 4.10 N 600 g 6.0 N 4.90 N 700 g 7.0 N 5.80 N 800 g 8.0 N 6.70 N 900 g 9.0 N 7.70 N 1000 g 10.0 N 8.30 N (N = newton)	Forcemeter Bucket of water Weights (100 g–1000 g)	The values in the table are only a guide. The values will depend on the weight and volume of the weights.
Sheet 23 (page 63) Direction of forces	2 The wind in the sail is pushing the boat forwards. 3 Gravity pulls the parachutist downwards. 4 Gravity pulls the skier down the slope. 5 The rugby player is being pushed to the left. 6 The snow plough is pushing the snow forwards. 7 The magnet pulls the paper clips to the left. 8 The horse pulls the sleigh to the right. 9 The discus thrower is pushing the discus to the left.	Toy car	In nature there are often several forces acting on an object at the same time. For example, the wind pushes the boat forwards but water resistance will be pushing against the motion of the boat.

Physical Processes

Forces and Motion

Activity	Answers	Resources needed	Teaching/ safety notes
Sheet 24 (page 64) Balanced forces	When I push upwards with my muscles, my body moves upwards. When I relax my muscles, my body hits the floor because of the force of gravity pulling me downwards. When my body remains in the same position, the pushing force from my muscles is balanced by the pulling force from gravity.		After a while the children will discover that, to keep themselves stationary, they have to continually exert a force upwards. This is more apparent the longer they keep pushing. They are only stationary when the force of gravity pulling downwards is balanced by the force from their muscles pushing upwards.
Sheet 25 (page 65) More balanced forces	• The recorded times will depend on the children themselves. • The weight of the boat (which is the pull of gravity on the boat) is balanced by the upthrust force from the water acting on the boat.	Timer	
Sheet 26 (page 66) Unbalanced forces	A force is required to make the ball speed up or slow down, change its shape and change its direction. As the ball is thrown upwards, the force of gravity pushes it downwards. When it falls and is caught, a pushing force is then used to push it back upwards.	Tennis ball	
Sheet 27 (page 67) Unbalanced forces puzzle	• Changing shape: egg being broken, concrete being broken up, paper being crushed. • Changing direction: snow boarder turning, satellite orbiting the Earth. • Changing speed: a discus being thrown, a space shuttle speeding up, a motorcycle speeding up.		

Physical Processes

Forces and Motion
Sheet 1

Push and pull forces

 A force is a **push** or **pull**. Look at these pictures.

A weightlifter pushing a weight upwards with a pushing force

A footballer pushing a ball forwards with a pushing force

A rower pulling an oar with a pulling force

 Look for pushing and pulling forces in these activities. Mark the pictures **push**, **pull** or **both**.

Forces and Motion Sheet 2 — Force words

 **Read** — We often use other words to describe pushing and pulling forces. For example, we sometimes say 'squash' to describe a pushing force which is used to press things.

 **Activity** — Look for five words which describe pull and push words in this word search.

t	w	i	s	t	d	s	r
e	y	l			m	q	c
a	a	x			t	u	g
r	n	w			o	e	v
f	k	q			h	e	k
d	r	a	g	w	b	z	s
s	q	u	a	s	h	e	v

 **Look further** — Find some more force words and write them below.

Forces and Motion Sheet 3

Magnetic forces

You need: 2 bar magnets and 2 horse-shoe magnets.

 Magnets **attract** some metal objects and they can also attract each other.

 Bring two magnets together and feel the magnetic force between them pushing and pulling at each other.

Magnets have a pole at each end called a **North pole** and a **South pole**. Bring each pole together as shown below. Put a tick in the correct box to show whether the magnets pushed apart or pulled together.

South — North	South — North	push apart ☐ pull together ☐
North — South	South — North	push apart ☐ pull together ☐
South — North	North — South	push apart ☐ pull together ☐

 Cross out the incorrect words.

When a North and South pole are brought together, we feel them **push/pull** each other **apart/together**. When two North or two South poles are brought together, we feel them **push/pull** each other **apart/together**.

Forces and Motion
Sheet 4

Magnetic materials

You need: a magnet and a selection of different materials to test.

Activity — Touch each of these objects with a magnet to find out if they are attracted to it. Use a selection of different objects if you wish.

 Another magnet	 A steel spoon	 A gold ring	 A plastic comb
 A wooden ruler	 A clay pot	 An iron bolt	 Rubber balloons
 Water	 Coins	 Paper clips	 A brick

Answer — Write down the names of the objects which were attracted to the magnet.

Physical Processes
44

© M Abraitis, A Deighan, B Gallagher, B Smith & M Toner
This page may be photocopied for use by the purchasing institution only.

Forces and Motion
Sheet 5

Magnetic poles

You need: a bar magnet and some paper clips.

 **Read**

The picture below illustrates how the **magnetic force** from a bar magnet can be used to attract a paper clip.

We cannot see this magnetic force but we can feel it when we hold a paper clip near to the ends (or poles) of a bar magnet.

 **Activity**

Hold a paper clip near to the ends of a bar magnet. What happens?

Use paper clips to find out where the magnetic force is strongest.

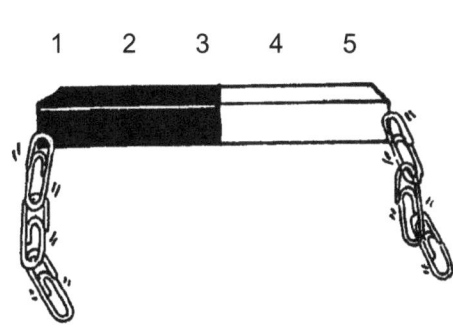

Part of magnet	Number of paper clips picked up
1	
2	
3	
4	
5	

 **Look further**

Which parts of the bar magnet are the strongest?

Forces and Motion
Sheet 6
Magnet shapes

You need: various magnets of different shapes, ruler and paper clips.

 Investigate which magnet will attract a paper clip from the furthest distance away.

Bar magnet

Horse-shoe magnet
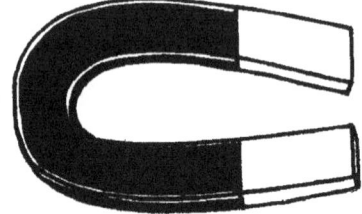

Ring magnet

Put a pile of paper clips on the table and hold each magnet in turn near the pile. The strongest magnet will pick up the most paper clips.

Which magnet is the strongest?

Forces and Motion
Sheet 7 — Attraction through materials

You need: a strong magnet, a paper clip and some different materials to test.

 Find out if a magnet can attract a paper clip through different materials. Tick the box if the magnet attracts the paper clip.

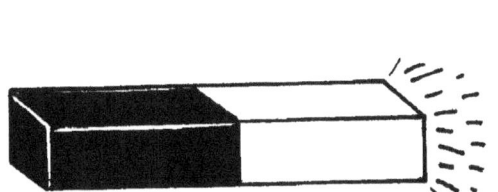

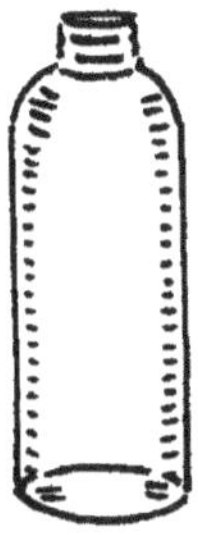 A plastic bottle ☐	 A school bag ☐	 An aluminium can ☐
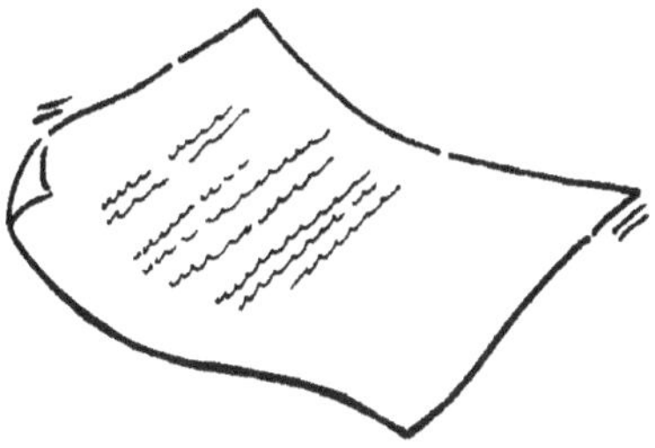 A sheet of paper ☐	 A steel cabinet ☐	 A bottle of water ☐
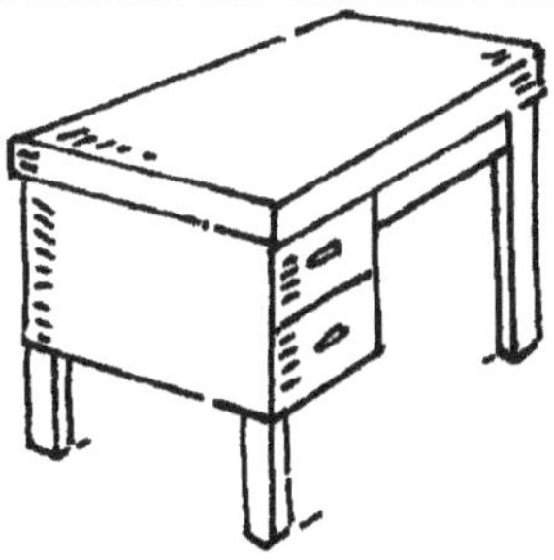 A wooden desk ☐	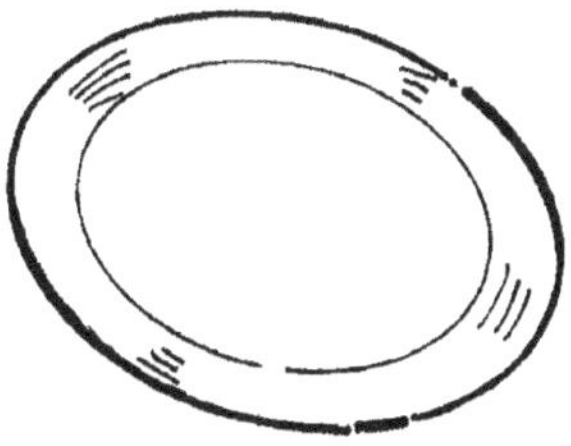 A dinner plate ☐	 A brick ☐

**Forces and Motion
Sheet 8**

The force of gravity

You need: a strong plastic bag, a large elastic band, string, a ruler and some objects of different sizes.

Gravity is a pulling force that pulls everything down towards the ground.

Gravity pulls everything downwards towards the ground.

Gravity pulls us to the Earth.

What could we do if we could switch off gravity?

Let's do an experiment to see the effect of gravity.

◆ Collect a ruler, a large elastic band, some string and a strong plastic bag.

◆ Use the string to tie the plastic bag to the elastic band.

◆ Place an object in the bag and lift it with the elastic band.

◆ Ask a partner to measure the length of the elastic band each time.

◆ Fill in the table below.

Object					
Elastic band length					

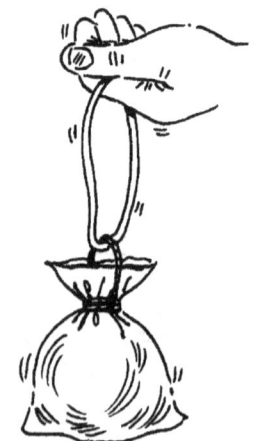

When objects are placed in the bag the elastic band stretches. This shows the force of gravity on each of the objects. The bigger the force (the heavier the object), the more the elastic stretches.

The force of friction

Forces and Motion — Sheet 9

Read

Friction is a force which is produced when two surfaces rub against each other. Sometimes friction can be useful and at other times it can be less useful.

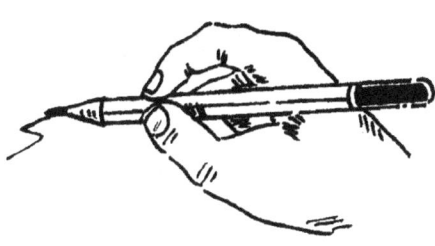

Friction is **useful** when you write on paper with a pencil because the force of friction between the pencil and paper rubs the graphite (lead) from your pencil onto the paper.

Friction can be **less useful** when you want to move down a slide quickly because the force of friction between your clothes and the slide holds you back and slows you down.

Activity

Look at the pictures below and decide if the friction caused between the surfaces rubbing against each other is useful or not. Tick the boxes if the friction is useful.

Wheels rubbing against the ground	The climber's shoes rubbing against the stone cliff	The tyres rubbing against the road	The moving parts of the engine rubbing against each other
Useful? ☐	Useful? ☐	Useful? ☐	Useful? ☐

Forces and Motion
Sheet 10 — Friction between different surfaces

You need: a large piece of card, a rubber, a plastic pen, a cardboard box and a soft toy.

Read

Friction is useful if we do not want to slip or fall when we are running or playing some types of games like netball or basketball.

These players wear rubber-soled shoes so that the friction between their shoes and the floor is great and they don't slip.

Activity

Carry out this experiment to find out which material produces the largest force of friction against cardboard.

Place a rubber, a plastic pen, a cardboard box and a soft toy on a piece of cardboard at the same height, then investigate what happens when the cardboard is slowly tilted.

Answer

Which material:
◆ slips down most easily?

◆ produces the greatest force of friction against the cardboard?

Forces and Motion
Sheet 11

Reducing friction

You need: a toy car, a toy truck and an elastic band/forcemeter.

 **Read**

Friction pulls at things which move and slows them down. You can see how friction works if you push a model car and let it roll across the floor.

The push force from you makes the car move.

The force of friction between the wheels and the floor slows the car down.

Many types of vehicles use wheels to reduce friction so that:
◆ they can go more quickly

◆ it takes a smaller pushing or pulling force to keep them going.

 **Activity**

Pull a large toy vehicle along the floor with an elastic band or a forcemeter.

Remove the wheels of the truck or jam them with paper and repeat the activity.

 **Answer**

It was easier to pull the truck with its wheels because...

If you used a forcemeter, fill in the table below.

Force needed to pull the truck with wheels	Force needed to pull the truck without wheels
_____ newtons	_____ newtons

© M Abraitis, A Deighan, B Gallagher, B Smith & M Toner
This page may be photocopied for use by the purchasing institution only.

Physical Processes

Forces and Motion
Sheet 12

Air resistance 1

You need: a feather and some stiff card.

 **Activity** Drop a feather and watch it carefully as it falls to the floor.

 **Answer** Draw the path of the feather as it falls to the floor.

 **Activity** Flap a piece of cardboard up and down. Watch the end of the cardboard carefully and feel what happens to the cardboard as it moves through the air.

 **Answer** Draw what you see and feel happening to the cardboard as it moves up and down.

Physical Processes © M Abraitis, A Deighan, B Gallagher, B Smith & M Toner
This page may be photocopied for use by the purchasing institution only.

Forces and Motion
Sheet 13

Air resistance 2

You need: 4 large pieces of paper or newspaper.

 **Activity** Let's find out if crumpling up a piece of paper affects how long it takes to fall to the floor.

◆ Drop a slightly crumpled piece of paper and a flat piece of paper to the floor.

◆ Now tightly crumple a piece of paper and drop it and a flat piece of paper again.

 **Answer** Did both pieces land on the floor at the same time?

Which landed on the floor first – the slightly crumpled piece or the flat piece of paper?

Which piece of paper fell fastest?

 **Look further** Think about what you have seen. Can you explain why the slightly crumpled ball of paper lands on the floor before the flat piece of paper and why the tightly crumpled piece falls fastest?

© M Abraitis, A Deighan, B Gallagher, B Smith & M Toner
This page may be photocopied for use by the purchasing institution only.

Forces and Motion
Sheet 14
Useful air resistance

You need: paper, scissors, thread, Plasticine and a stop clock or watch.

Activity Look at these two pictures and write down why each of them shows air resistance being useful.

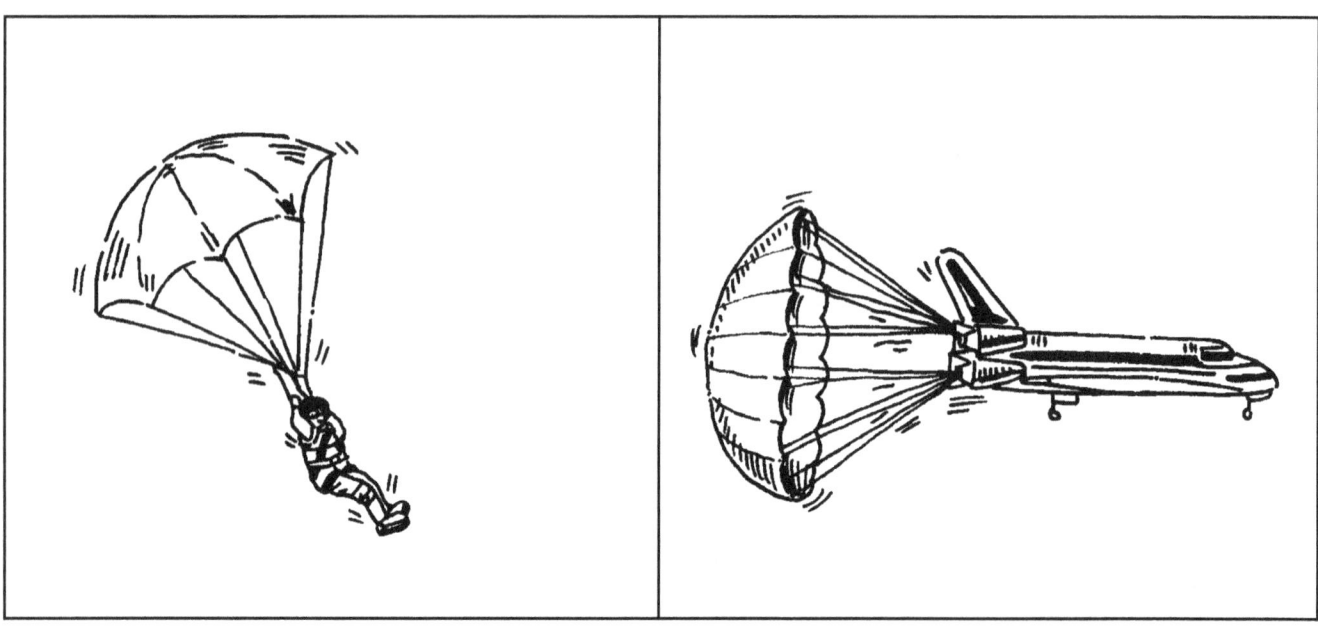

◆ Make your own parachutes using the different shapes below.

◆ Time how long each takes to reach the floor. Which shape is the best one for slowing down the parachute?

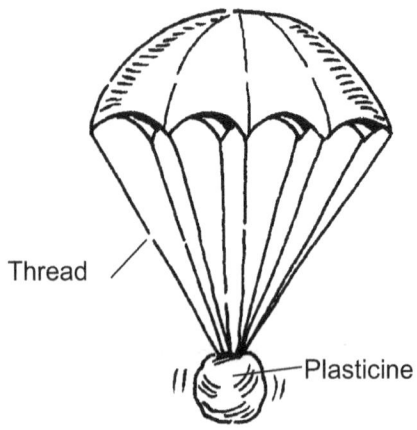

Shape chosen	Circle	Square	Rectangle	Your own shape
Time taken to hit floor				

Answer The _____ shape makes the best parachute.

Physical Processes
54

© M Abraitis, A Deighan, B Gallagher, B Smith & M Toner
This page may be photocopied for use by the purchasing institution only.

Forces and Motion
Sheet 15

Streamlining 1

You need: scissors and glue.

 **Read**

The pictures below show people who are trying to move very quickly. They are crouching down so that air resistance does not slow them down too much.

Changing shape to move quickly is called **streamlining**. Some things are made streamlined and some things are born streamlined.

 **Activity**

Cut out the animals below and order them into streamlined and not streamlined groups (streamlined animals can move very quickly).

© M Abraitis, A Deighan, B Gallagher, B Smith & M Toner
This page may be photocopied for use by the purchasing institution only.

Physical Processes

Forces and Motion
Sheet 16

Streamlining 2

 **Read**   The vehicles below are all streamlined.

 **Answer**   What do you think makes the vehicles streamlined?

 **Look further**   Design and draw your own streamlined vehicle.

Physical Processes   © M Abraitis, A Deighan, B Gallagher, B Smith & M Toner
This page may be photocopied for use by the purchasing institution only.

Forces and Motion Sheet 17: The spinner experiment

You need: scissors, a sheet of A4 paper, paper clips, a metre stick and a timer.

 First make a spinner from a piece of A4 paper. Fold along full lines in the direction of the arrows and cut along the dotted lines.

Plan an investigation to find out which factors affect how long it takes for the spinner to reach the floor after being dropped.

Investigate:
- ◆ the height from which it is dropped
- ◆ the length of its wings
- ◆ the number of paper clips hanging from it.

 Draw a table for your results.

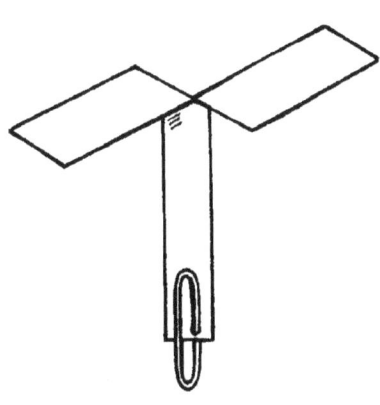

**Forces and Motion
Sheet 18**

Water resistance

You need: some Plasticine, a timer, string, wallpaper paste and a large cylinder.

 If you have ever tried to walk in the sea when it is up to your waist you will have noticed that it is difficult to move quickly. This is because there is **friction** between you and the water. This friction is called water resistance.

 Let's investigate how water resistance affects different shapes of Plasticine.

Time how long it takes for the three shapes to fall to the bottom. You could thicken the water with wallpaper paste to increase the time taken for them to fall to the bottom.

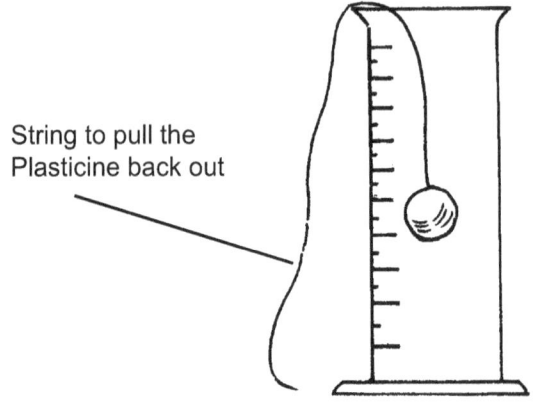

String to pull the Plasticine back out

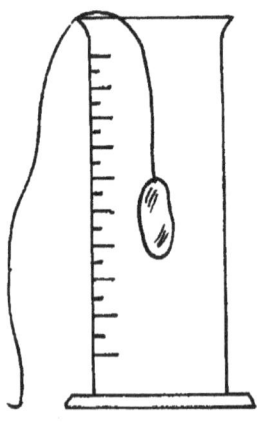
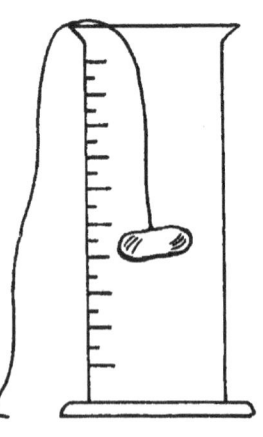

time ___ seconds   time ___ seconds   time ___ seconds

 Think about different shapes to try. What things are you going to keep the same to make the test fair?

Forces and Motion Sheet 19: Springs and elastic

Read: When an elastic band is **stretched** it will exert a pulling force on the object to which it is attached. The more the elastic band is stretched, the greater the pulling force exerted by the elastic band will be. When a spring is stretched or **compressed**, it also exerts a force on whatever is stretching or compressing it.

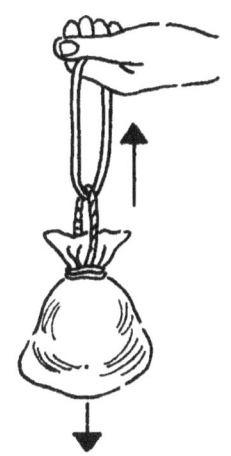

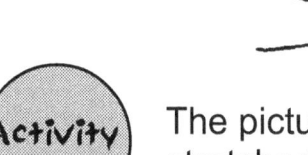

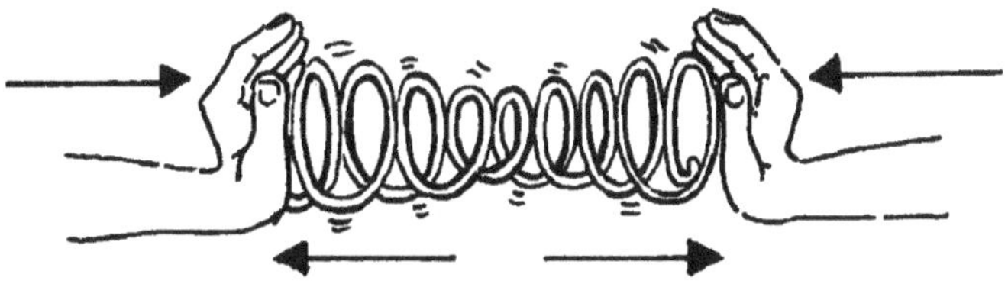

Activity: The pictures below show how sports people can use the forces from stretched or compressed springs and elastic bands. Look at the pictures and write down how the forces are being used.

1. Bow and arrow
2. Sling shot
3. Pogo stick
4. Spring board
5. Trampoline
6. Bungee rope

Stretched elastic

**Forces and Motion
Sheet 20**

You need: a toy car, a ruler, board with 2 screws in it and a large elastic band.

 **Activity** Investigate what effect stretching the elastic band has on the speed of the car and the distance it travels.

◆ Place the elastic band round the screws in the board.

◆ Pull it back, and place the toy car in the bow.

◆ Let go of the band and see how far the car travels along the board.

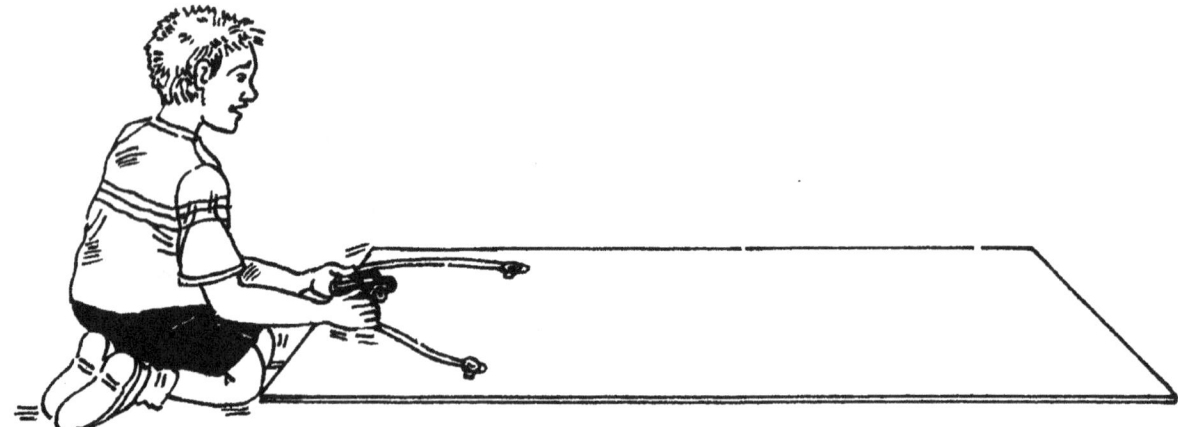

Length the elastic is stretched				
Distance travelled by the car				

 **Answer** Describe what effect stretching the elastic band has on the motion of the car.

Forces and Motion
Sheet 21 — The forcemeter

You need: a forcemeter, string and some materials for testing.

A **forcemeter** is used to measure the size of forces. The size of a force is measured in **newtons**.

◆ Examine the forcemeter and look for the spring.

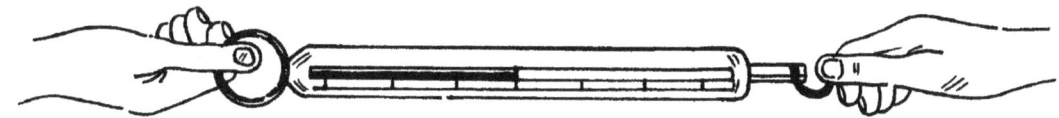

◆ Hold the forcemeter while your partner gently pulls the other end.

◆ Look to see what happens to the spring in the forcemeter when your partner pulls it.

◆ Note down the reading on the forcemeter when you try these activities. Tie the books to the forcemeter with string.

Opening a drawer takes a force of _____ newtons.	Pulling some books across the table takes a force of _____ newtons.	Loosening a shoe lace takes a force of _____ newtons.	Lifting a jacket takes a force of _____ newtons.

Forces and Motion — Sheet 22
Upthrust

You need: a forcemeter, a bucket or basin of water and 100 g to 1000 g weights.

 **Activity** Let's find out if an object weighs the same in water as it does in air.

Weigh the object with a forcemeter in air, record its weight and then submerge the object in water and record its weight again.

Mass of the object	Weight of object in air	Weight of object in water
100 g	newtons	newtons
200 g	newtons	newtons
300 g	newtons	newtons
400 g	newtons	newtons
500 g	newtons	newtons
600 g	newtons	newtons
700 g	newtons	newtons
800 g	newtons	newtons
900 g	newtons	newtons
1000 g	newtons	newtons

 **Look further** This elephant seal weighs so much that it can hardly hold itself up when it is out of water. Find out what force helps keep it afloat when it is in water.

Forces and Motion
Sheet 23

Direction of forces

You need: a toy car.

 **Read** When something is pushed or pulled with enough force then it will move in the direction of the push or pull force.

Try pushing a toy car along the floor.

Direction of force   Direction of car

Forces always act in a particular direction.

 **Activity** Look at the pictures below and draw with arrows the direction of one of the forces. The first one is done for you.

1 — The weightlifter pushes the weights upwards	2	3
4	5	6
7	8	9

Forces and Motion
Sheet 24

Balanced forces

Activity Try this experiment:
◆ Find a clear space on the floor.

◆ Try doing five push-ups as slowly as you can.

◆ Think about which muscles you are using to push up with and what you are pushing against.

◆ Repeat the push-ups several times and, when you have your breath back, complete the sentences below.

Answer

When I push upwards with my _____ , my body moves _____ . When I relax my _____ , my body hits the floor because of the force of _____ pulling me _____ .

Activity When you did your push-ups, you pushed upwards with a force greater than the force of gravity which was pulling you downwards, so your body moved upwards. When the force of gravity pulling you downwards was greater, you moved downwards.

Now repeat the press-ups, but this time stop half-way up and keep your body in that position for as long as you can.

Pushing force from muscles   Pushing force from muscles

Answer

When my body remains in the same position, the _____ force from my _____ is balanced by the pulling _____ from _____ .

Pulling force of gravity

Physical Processes
64

**Forces and Motion
Sheet 25**

More balanced forces

You need: a timer.

 **Read** — When the forces acting on an object which is not moving are balanced, then the object will stay still.

 **Activity** — Time how long you can keep the pushing force of your muscles balanced against the pulling force of gravity. Remember, the forces are only balanced if you do not move up or down.

  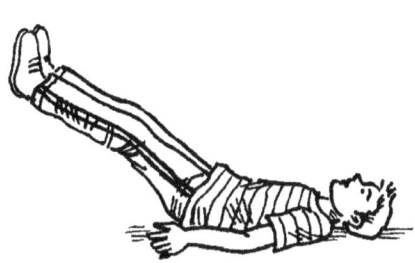

I kept the forces balanced for...	I kept the forces balanced for...	I kept the forces balanced for...

 **Look further** — Can you describe why a boat does not sink because of gravity pulling it downwards?

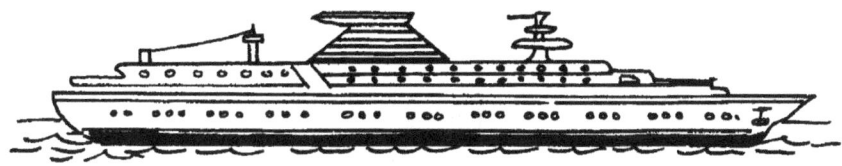

Forces and Motion
Sheet 26

Unbalanced forces

You need: a tennis ball.

Try the following three activities with the ball.

◆ Make the ball speed up by pushing it along the floor.

◆ Change the ball's shape by squeezing it in your hand.

◆ Make the ball change direction by throwing it upwards and catching it.

What is needed to change the speed, shape or direction of different objects?

Forces and Motion Sheet 27
Unbalanced forces puzzle

Activity: Draw lines to match the pictures of unbalanced forces with the correct captions.

| An unbalanced force changing something's shape | An unbalanced force changing something's direction | An unbalanced force changing something's speed |

A discus being thrown

A snow boarder turning

An egg being broken

A satellite orbiting the Earth

A space shuttle speeding up

Paper being crushed

A motorcycle speeding up

Concrete being broken up

© M Abraitis, A Deighan, B Gallagher, B Smith & M Toner

Light and Sound

Background information

Everyday effects of light

Light travels from a source such as a flame in straight lines and at the fantastic speed of around 300 million metres per second.

Shadows are created behind objects, such as sticks, which do not allow light to pass through them. These types of materials are called **opaque** objects. Materials which allow all or a great deal of light to pass through them are called **transparent** materials. Materials such as grease-proof paper which allow some of the light to pass through them are called **translucent** materials.

Almost all materials reflect some light from their surface (except for purely black materials). Highly polished materials and mirrors are very good at reflecting light from their surfaces but dull or rough surfaces are not. Objects which reflect light can appear coloured because they reflect only some of the light which shines on them. For example, grass appears green in sunlight because it mostly reflects the green coloured light within sunlight. The rest of the colours are absorbed by the grass.

Seeing

We see things because light from them, either emitted by them or reflected from them, enters our eyes. This light energy entering our eyes is changed into an electrical signal by nerve cells on the retina. The electrical signals are converted into an image by the brain.

Helpful words
- Source
- Shadow
- Rotation
- Sundial
- Opaque
- Transparent
- Translucent
- Reflection
- Vibration
- Pitch
- Loudness

Vibration and sound

When an object shakes or vibrates in air it causes vibrations in the air which travel outwards at around 300 metres per second. We hear these sounds because the air vibrations set up similar vibrations in the ear.

Musical instruments produce sounds by the use of vibrating strings (as in a guitar), vibrating surfaces (such as drums or cymbals), and vibrating air columns (as in a flute). The loudness of these sounds depends on the height (**amplitude**) of these vibrations. The larger the amplitude of the vibration, as in hitting a drum skin with a large force, then the louder the produced sound will be.

The **pitch** of sound depends on the number of vibrations per second. The more vibrations there are per second, the higher the pitch of the sound. The pitch of a drum skin depends on the tightness of the drum skin, and the pitch of a stringed instrument depends on the length of the string, the thickness of the string and the tightness of the string. The pitch of a wind instrument depends on the length of the vibrating air column (which can be changed by covering up the holes).

Light and Sound

Sound vibrations travel through gases (such as air), liquids (such as water) and solids (such as walls). Sound will travel through different materials at varying speeds. It travels most quickly through solids, less quickly through liquids and is slowest through gases such as air. (Air is a mixture of gases and is made up mostly of nitrogen.) The loudness of sound is measured in **decibels** (db). Loud sounds can damage hearing permanently. It is important for people who work in noisy environments to protect their ears from permanent damage by wearing ear protectors. These work by placing a material between the ear and the source of the noise to absorb some of the sound energy.

Some ideas for further work

◆ Design and build a water or candle clock. How could the clock be calibrated to tell the time accurately?

◆ Find out about modern glass used in fibre optics (cable TV, the Internet, etc).

◆ Investigate which type of clothing is best at reflecting light at night. Torch beams could be shone on different luminous clothing in the dark.

◆ Listen to tapes of recorded 'whale songs'. Research into why whales use sound in this way.

◆ Devise you own musical instruments from everyday items. Think about things you could use or adapt, for example hollow tubes, cardboard boxes, string, sticks, bottles, etc. Think about how some of the instruments could be 'tuned'. Practise a simple tune as an 'orchestra' and give a class performance.

◆ The volume of sound is measured in decibels. What is the decibel scale and how many decibels does it take to damage your hearing?

Light and Sound

Activity	Answers	Resources needed	Teaching/safety notes
Sheet 1 (page 74) A light in the dark	When the light is blocked by a piece of card a shadow is formed. Light is not able to bend around the card because it travels in straight lines. **Sources of light** 3 A burning piece of wood 6 An oil lamp 1 A wax candle 5 A gas light 2 An electric bulb 4 A low-energy electric bulb	Torch Scissors Dark card	Never look directly at the Sun or bright lights. Cutting teeth into the card may help to show that light travels in straight lines. We require sources of light to see in the dark.
Sheet 2 (page 75) How shadows are formed	The shadows should be the same as the shapes on the projector.	Card Scissors Overhead projector Cloth Plastic bag	Shadows are similar in shape to the objects which form them.
Sheet 3 (page 76) Shadow shapes	1 matches c 2 matches a 3 matches d 4 matches b		Children could be given some different shaped objects and asked to draw their shadows.
Sheet 4 (page 77) Materials and light 1	• Transparent: glass. • Opaque: wood, brown paper. • Translucent: nylon shirt, bubble wrap, tissue, grease-proof paper.	Different materials	Water and cling film are other examples of transparent materials.
Sheet 5 (page 78) Materials and light 2	• Complete shadow: wood, brown paper. • Some shadow: nylon shirt, bubble wrap, tissue, grease-proof paper. • No shadow: glass.	Different materials Overhead projector	
Sheet 6 (page 79) Reflection of light	As the angle between the mirror and the light beam changes then the angle of the reflected beam also changes. These two angles are equal.	Torch Hand mirror Flat plastic mirror Paper Light box Glue stick	The ray of light is more easily seen if the room lights are switched off. The reflected ray of light from the light box is easier to see if the flat mirror is tilted slightly towards the paper.

Physical Processes

Light and Sound

Activity	Answers	Resources needed	Teaching/ safety notes
Sheet 7 (page 80) Useful mirrors	• Car mirrors reflect light coming from behind the car so that drivers can see what is behind them. • The soldier's periscope allows him to see what is over the wall. • The mirror behind the torch bulb reflects the light from the bulb forward as a beam of light.	Ruler Protractor 2 small plastic mirrors Glue stick	When making a periscope it is important to make sure that the two mirrors are angled as near to 45 degrees as possible.
Sheet 8 (page 81) Reflecting light to be safe	Shiny and smooth surfaces reflect light better than dull and rough surfaces.	Torch Materials to test	Modern fibre optics used in cable television and communication networks rely on light being reflected down a glass cable for them to work.
Sheet 9 (page 82) Sources of light	• Source of their own light: log fire, television picture, lit bulb, match flame, Sun. • Reflect light in order to be seen: books, cloud, people, tree.	Scissors Glue	
Sheet 10 (page 83) Our eyes	• Left, top to bottom: Cornea, Lens, Pupil. • Right, top to bottom: Retina, Optic nerve.	Scissors Glue Reference materials	Commercially available CD-ROMs contain excellent graphics of the eye.
Sheet 11 (page 84) Making sounds	• The harder the ruler is flicked, then the bigger the amplitude of vibration and the louder the sound will be. • The pitch (frequency) of the sound produced changes as the length of the ruler hanging over the table changes. • The pitch (frequency) depends on the tightness of the elastic band.	Plastic ruler (unbreakable) Elastic band	Safety: Use plastic unbreakable rulers. When investigating the relationship between the loudness of the sound and the size of the amplitude, it is important to keep constant the length of the ruler extending over the table.

Physical Processes

Light and Sound

Activity	Answers	Resources needed	Teaching/ safety notes
Sheet 12 (page 85) Seeing sound vibrations	• It is hard to see the vibrations of a tuning fork but when it is placed in the water, the water shakes, showing that the tuning fork is indeed moving (vibrating) to produce the sound. • When the drum skin which has pasta shells on it is hit, the pasta shells jump around. This shows that the drum skin moves (vibrates) to produce the sound.	Tuning fork Water holder Dry pasta shells Drum	Different tuning forks are designed to vibrate at different frequencies to produce different notes. Try different tuning forks in the water to investigate whether they affect the water differently.
Sheet 13 (page 86) Types of sound	The type of sound heard will depend on how hard the object is hit and from what the object is made. Hard materials should produce high-pitched sounds and soft materials should produce low-pitched sounds.		You could look at different musical instruments such as drums, cymbals, xylophones and triangles and write down the materials used to make them.
Sheet 14 (page 87) Sound pitch	It is the air column which vibrates to produce the sound. The smaller the air column then the higher the pitch of the sound produced. The air inside the bottles produces the sound when struck. Hence the difference in pitch.	Small wooden mallet Empty glass bottles	A pencil could be used if a wooden mallet is not available. The bottles should be identical but tall glasses could be used instead of bottles. Take care using glass bottles.
Sheet 15 (page 88) Pitch and loudness	• Guitar: the pitch is varied by the different lengths, thickness, and tightness of the strings. • Pan pipes: the pitch is varied by the different lengths of air columns in the pipes. • Drum: the pitch is varied by the tightness and size of the drum skin. • Recorder: the pitch is varied by covering up holes which vary the length of the air column in the pipe. • Xylophone: the pitch is varied by changing the length of the bars.	Guitar Pan pipes Drum Recorder Xylophone	

Light and Sound

Activity	Answers	Resources needed	Teaching/ safety notes
Sheet 16 (page 89) Does sound travel through air?	We hear these sounds because sound can travel through air.		This activity is intended to focus on the fact the sound travels through the air and that all sounds (loud, soft, high and low) travel just as easily through air.
Sheet 17 (page 90) Does sound travel through solids?	• The sound from the clock should be heard through the stick, demonstrating that sound travels through solids. • String is also a solid and the string telephone is another fun way of demonstrating that sound travels through solids.	Clockwork clock Piece of wood 2 plastic cups Ball of string	The string telephone has to be held quite tightly in order to work, so it is a good idea to use a strong piece of string. Make sure children do not poke their ears with the stick of wood.
Sheet 18 (page 91) Does sound travel through liquids?	• When the toy is squeezed, sound can be heard. This demonstrates that sound can travel through liquids such as water. • Dolphins and whales communicate by sound. They also use sound echoes to locate fish. Killer whales can stun fish with sound.	Squeaky toy Basin of water	Sound travels very quickly through water (four times as fast as in air). This helps whales and dolphins to communicate over very large distances.
Sheet 19 (page 92) Stopping sound	Some of the sound is absorbed by the materials so that not all of it reaches our ears.	Clockwork clock Bubble wrap Woollen scarf Foam sheets	

Physical Processes

A light in the dark

Light and Sound Sheet 1

Warning: **Never** look directly at the Sun or bright lights.

You need: a torch, scissors and a piece of dark card.

 **Activity** Shine a torch against a wall. You will see the light against the wall. This shows that light travels. Try the experiment in a darkened room – it works better and is much more fun!

 **Answer** Describe what happens when you block the light beam with a piece of dark card.

 **Activity** From prehistoric times, people have had to use other sources of light at night. Arrange the sources of light below, starting with the oldest one. Think about how sources of light developed.

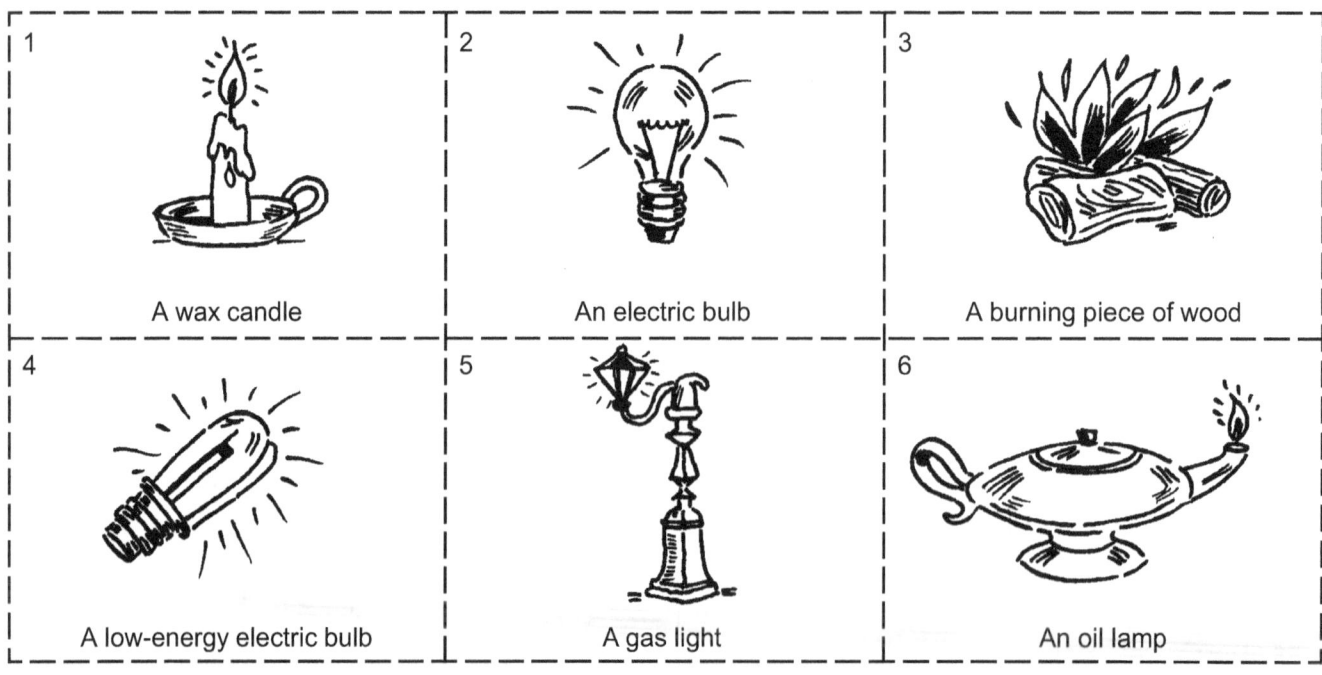

1	2	3
A wax candle	An electric bulb	A burning piece of wood
4	5	6
A low-energy electric bulb	A gas light	An oil lamp

Physical Processes
© M Abraitis, A Deighan, B Gallagher, B Smith & M Toner

Light and Sound Sheet 2: How shadows are formed

You need: some card, scissors, an overhead projector, some cloth and a plastic bag.

 **Activity** Make card shapes and put them on the projector. Draw the shadow they make.

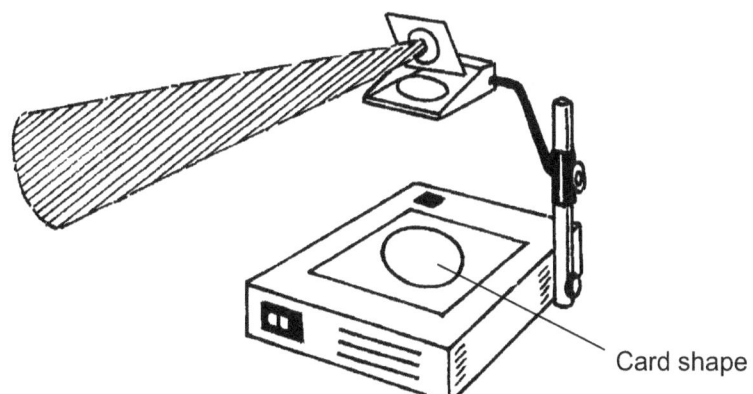

Card shape

Card shape	▬	▲	✚	●
Shape of shadow formed on the wall				

 **Read** Shadows are formed when light cannot pass through an object. The shape of a shadow depends on the shape of the object.

 **Look further** Try the activity again, but this time use shapes made out of different materials, such as cloth and plastic.

Light and Sound
Sheet 3

Shadow shapes

The silly people below have got their shadows mixed up. Can you find each person's shadow for them? Draw a line between each person and their shadow.

1    a

2    b

3    c

4    d

Materials and light 1

Light and Sound — Sheet 4

You need: lots of different materials (see below).

- Materials which we can see through are called **transparent** materials.
- Materials which we cannot see through are called **opaque** materials.
- Materials in between are called **translucent** materials.

Test the materials below by holding them up to a window.

Material	Transparent (*All* the light passes through)	Opaque (*None* of the light passes through)	Translucent (*Some* of the light passes through)
Wooden ruler			
Nylon shirt			
Bubble wrap			
Glass cup			
Grease-proof paper			
Brown paper bag			
Tissue paper			

Materials and light 2

Light and Sound Sheet 5

You need: lots of different materials (see below) and an overhead projector.

Activity
- ◆ Use an overhead projector to investigate which type of material makes the best shadow.
- ◆ Try other materials from around the room.

Material	Complete shadow	Some shadow	No shadow
Wooden ruler			
Nylon shirt			
Bubble wrap			
Glass cup			
Grease-proof paper			
Brown paper bag			
Tissue paper			

Look further Which materials make the best shadows?

Physical Processes

© M Abraitis, A Deighan, B Gallagher, B Smith & M Toner
This page may be photocopied for use by the purchasing institution only.

Reflection of light

Light and Sound
Sheet 6

You need: a torch, a hand mirror, a flat plastic mirror, paper, a light box and a glue stick.

Read

We cannot see round corners because light travels in straight lines. But we can use mirrors to reflect light in different directions.

Activity

◆ Make your room as dark as you can.

◆ Use a mirror to reflect light from a torch onto the walls and ceiling of your room.

◆ Now change the angle at which you are holding the mirrors, and watch what happens to the torch beam.

We can draw the rays of light by sticking a mirror onto a piece of paper.

◆ Stick a flat plastic mirror onto a sheet of white paper near one end. Fold the paper so that the mirror is at an angle to the rest of the paper.

◆ Shine a ray of light from the light box onto the mirror at different angles.

◆ Record what happens to the reflected beam by drawing over the lines of light with a pencil.

Light and Sound
Sheet 7

Useful mirrors

You need: a ruler, a protractor, 2 small plastic mirrors and a glue stick.

 **Activity** Find the useful mirrors in these three pictures.

A car	A soldier's periscope	A torch

 **Activity** Let's make a periscope.

♦ Using a protractor, mark 45 degree lines at the top and near the bottom of a ruler.

♦ Stick mirrors on the lines you have marked. Make sure the mirrors have their shiny sides facing each other.

♦ Hold the ruler below the bottom mirror and look into that mirror.

♦ You should be able to see a reflection from the top mirror.

Light and Sound Sheet 8 — Reflecting light to be safe

You need: a torch and some materials to test – see below.

Read — When we are out cycling it is important that we are safe. We can help make ourselves safe by wearing bright clothes and reflective safety bands so that others can see us, especially if it is dark.

This cyclist helps keep safe by wearing brightly coloured clothes and having special light reflectors on the back and side of his bike.

Activity — Investigate which type of material is best at reflecting light by:
a) looking if you can see your face in it, and
b) testing if it reflects a torch beam.

Material	Can you see your face in it?	Does it reflect a torch beam?
Mirror		
Perspex		
Piece of dull wood		
Piece of polished wood		
Gloss painted radiator		
Matt painted wall		
Bike reflector		
Jumper		
The surface of a basin of water		
Cooking foil		
Paper		

© M Abraitis, A Deighan, B Gallagher, B Smith & M Toner

Sources of light

Light and Sound
Sheet 9

You need: scissors and glue.

 **Read** We can see things because light from them enters our eyes. Some things, like a lamp, are sources of their own light. Other things, like this page, are seen only because light from a source reflects off them and then enters our eyes.

 **Activity** Cut out the pictures and arrange them into things which are sources of their own light and things which are seen because they reflect light. (Hint! Which objects could you still see in the dark?)

1 A log fire	2 Some books	3 A television picture
4 A cloud	5 A lit bulb	6 Us
7 A match flame	8 The Sun	9 A tree

Physical Processes

© M Abraitis, A Deighan, B Gallagher, B Smith & M Toner
This page may be photocopied for use by the purchasing institution only.

Light and Sound
Sheet 10

Our eyes

You need: scissors and glue.

 We can see things only when light enters our eyes. Our eyes are very important.

 The diagram below shows a cut-through eye. Use books or a CD-ROM to find information to help you match the labels with the correct part of the eye.

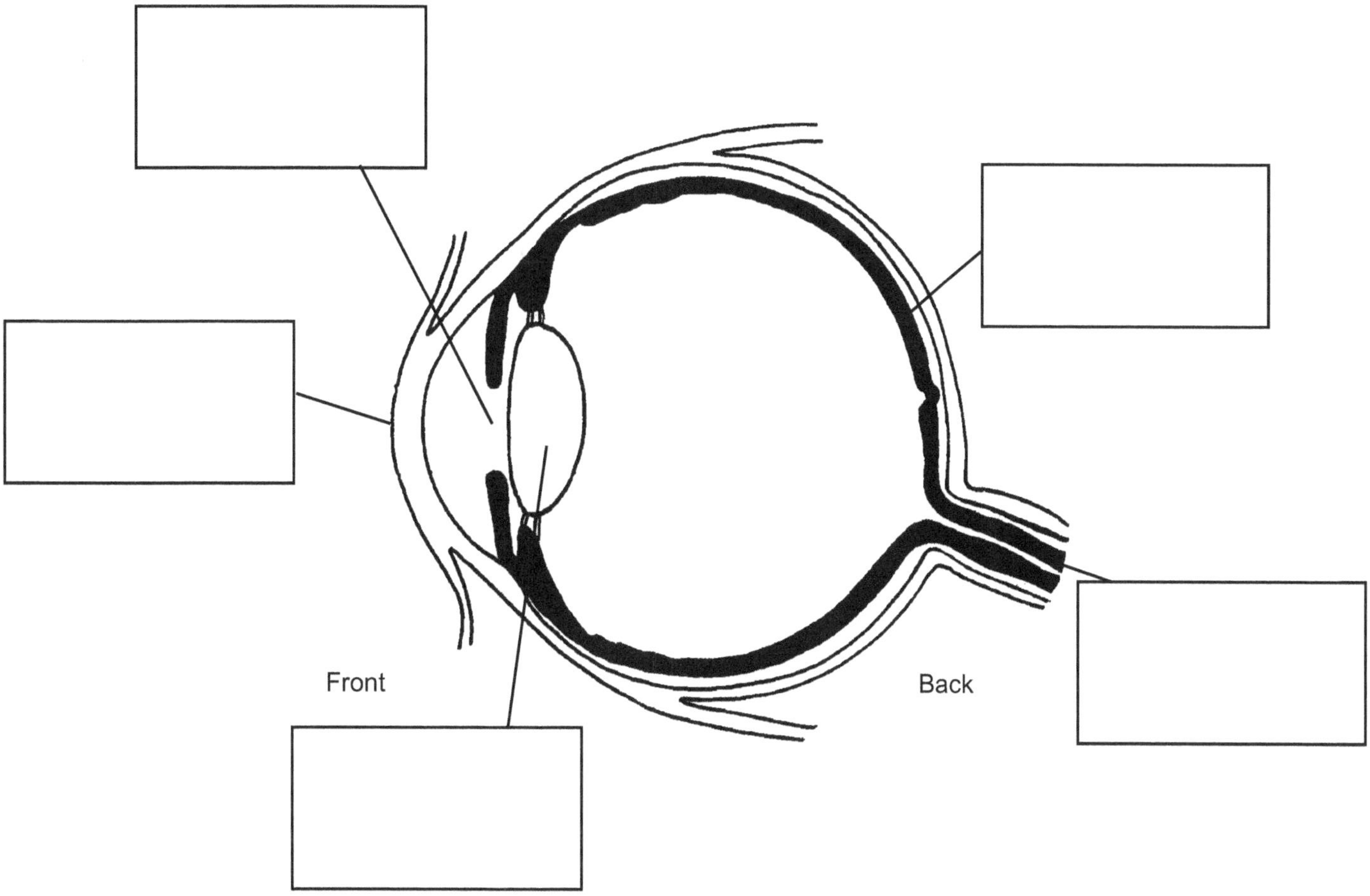

Front   Back

Lens, which focuses the light into an image at the back of the eye

Retina, which changes light to an electrical signal

Optic nerve, which carries the electrical signal to the brain

Cornea, the transparent outside of the eye

Pupil, which allows light in

Light and Sound
Sheet 11

Making sounds

You need: a plastic unbreakable ruler and an elastic band.

Sound is made when something is made to **vibrate** (or shake very quickly).

◆ Make sound by holding one end of a ruler tightly on a desk and flicking the other end quickly with your other hand.

◆ Now change the sound by changing the length of the ruler hanging over the desk.

Does the sound get higher or lower as the ruler gets shorter?

◆ Stretch an elastic band across the fingers of one hand and pluck it with the other hand.

◆ Now change the sound by changing how much you stretch the elastic.

Does the sound get higher or lower as the elastic band gets looser?

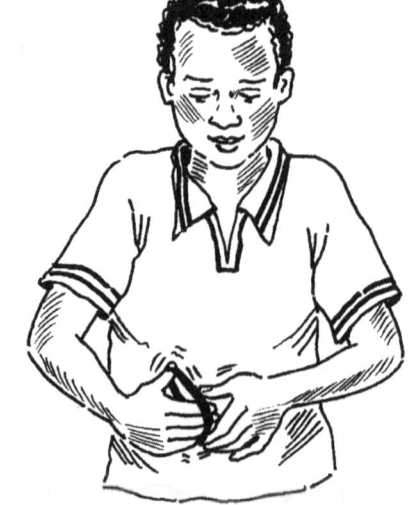

Seeing sound vibrations

Light and Sound
Sheet 12

You need: a tuning fork, a water holder, some dry pasta shells and a drum.

◆ Tap a tuning fork on a hard surface to make it vibrate and produce a sound.

◆ While the tuning fork is still making a noise, place the top of it onto the surface of a jug of water.

Describe what happens to the surface of the water when the turning fork touches it.

◆ Place some pasta shells on the skin of a drum.

◆ Tap the drum.

Describe what happens to the pasta shells.

Why does this happen?

Light and Sound
Sheet 13

Types of sound

 **Activity**

Sounds are made when things vibrate (shake very quickly) but what causes the different types of sounds?

◆ Use a pencil to tap each of the objects below and listen to the sound they make.

◆ Complete the table, saying whether you thought the sound was **tinny** (high pitched), **dull** (low pitched), or **hollow** (in between).

Object	Is the object hard or soft?	Describe the sound you hear
Shoe		
Bottle		
Can		
Chair		
Filing cabinet		

 **Read**

The type of sound you hear when you hit something depends on the material from which it is made. Hard materials normally give a high-pitched sound and soft materials normally give a low-pitched sound.

Sound pitch

Light and Sound Sheet 14

You need: a small wooden mallet and some empty glass bottles.

 Pitch is how high or low a sound is. Investigate changing pitch.

Fill some bottles with different amounts of water and hit them softly with a small wooden mallet to hear the pitch change.

◆ Find out how to produce high and low pitch sounds.

◆ Tune some bottles so that you can play a song, such as 'Three Blind Mice'.

 Now try blowing across the tops of the bottles to produce sound.

 Which musical instruments produce sounds in a similar way?

What vibrates to make the sound when you hit or blow across the bottles?

What is interesting about the difference in the sound created by tapping and then blowing across the same bottle?

Pitch and loudness

Light and Sound — Sheet 15

You need: a guitar, Pan pipes, a drum, a recorder and a xylophone.

Activity: If you have the musical instruments below, find out how to change the pitch and loudness of each instrument. Write down what you discover. (Hint! Look for changes in length, thickness and tightness.)

A guitar	
Pan pipes	
A drum	
A recorder	
A xylophone	

Physical Processes
88

© M Abraitis, A Deighan, B Gallagher, B Smith & M Toner
This page may be photocopied for use by the purchasing institution only.

Light and Sound
Sheet 16

Does sound travel through air?

Read

When the school bell rings we hear it because the bell makes the air vibrate. The vibrations travel through the air to our ears. These vibrations in the air are called sounds.

Activity

The pictures below show how other sounds are produced in the air. Describe what the sound is like for each one.

Useful words: loud, booming, buzzing, high-pitched, shrieking, howling.

	An aeroplane taking off
	A flying insect
	A wolf calling out

Light and Sound Sheet 17: Does sound travel through solids?

You need: a ticking clock, a rounded piece of wood (30 cm long), 2 plastic cups and a ball of string.

Read — If you have ever heard someone speak behind a closed door then you will know that sound travels through solid objects.

Activity — Try these two ways of showing how sound travels through solids.

◆ Hold a piece of wood to a ticking clock and listen carefully to the other end. What do you hear?

◆ Let's make a string telephone.

1. Make a small hole in the bottom of two soft plastic cups, thread some string through the holes and tie big knots on the ends.

2. Ask a partner to put one cup to his or her ear.

3. Test the telephone by stretching the string and speaking into the other cup. The string must be really tight.

Light and Sound Sheet 18: Does sound travel through liquids?

You need: a squeaky toy and a basin of water.

Activity — Find out if sound travels through water by squeezing a squeaky toy in a bowl of water. Soak the toy completely in the water. Then squeeze the toy.

Answer — Does the sound from the toy travel through the water?

Look further — Whales and dolphins both live in the water. Find out how these two animals use sound.

Light and Sound
Sheet 19

Stopping sound

You need: a clockwork clock, some bubble wrap, a woollen scarf and some foam sheets.

Read

Loud sounds can damage our hearing and cause deafness.

The man in the picture is wearing ear protectors over his ears to protect them from the loud sounds made by the drill.

Activity

Investigate which materials are the best at muffling sound. Cover a clock with each of the materials and then listen to find out how good the material is at muffling the sound.

Material	The material muffles *all* the sound	The material muffles *some* of the sound	The material muffles *none* of the sound
Bubble wrap			
Wool			
Foam sheeting			

Look further

Plan an investigation to find out if the thickness of a material affects how well the material muffles sound. Try using the materials you used in the last activity.

The Earth and Beyond

Background information

The Sun is a star at the centre of our solar system. It is thought that the Sun is around 4700 million years old and that it will continue in its present form for another 5000 million years. Then it will expand and engulf the inner planets before it shrinks to become a white dwarf star.

The Sun is a giant nuclear reactor at the centre of which **hydrogen** is compressed to form **helium** which results in the release of energy. This energy causes the Sun's inner temperature to reach 15 million °C and its surface temperature to be around 5550 °C. The Sun is the source of all the energy available on Earth. It is the Sun's heat which warms the Earth's surface and controls the climate. The light from the Sun is converted into food by plants which ultimately feed all life on Earth.

Nine planets in the solar system have been observed, although some astronomers believe that there may be a tenth planet. It is thought that the nine known planets were created from a swirling cloud of dust around 4600 million years ago. This helps explain why the nine planets all travel anti-clockwise around the Sun. The particles which made up the dust cloud were the remains of stars like our Sun which had existed long ago and had eventually blown apart as supernovae (the universe is thought to be between 10,000 million and 20,000 million years old).

The Sun, planets and most moons are almost spheres because, as they were forming out of the dust cloud, they were soft and gaseous. The force of gravity was able to pull them into a spherical shape. As the Earth formed from the dust cloud, it was like Jupiter and very hot. As the Earth cooled, it solidified on the outside but remained molten in the middle.

Helpful words
- Space
- Planet
- Orbit
- Moon
- Sun
- Rotation
- Phases

Periodic changes

The planets still retain some of the movement energy (kinetic energy) from the original dust cloud from which they were formed. This is apparent in their motion around the Sun, their own rotation and in the motion of the moons which orbit the planets.

This motion, together with the tilt and rotation of the Earth, gives rise to day and night and the four seasons. The Earth spins on its own axis approximately every 24 hours, takes approximately 365 days to orbit the Sun and has a moon which takes approximately 28 days to orbit the Earth.

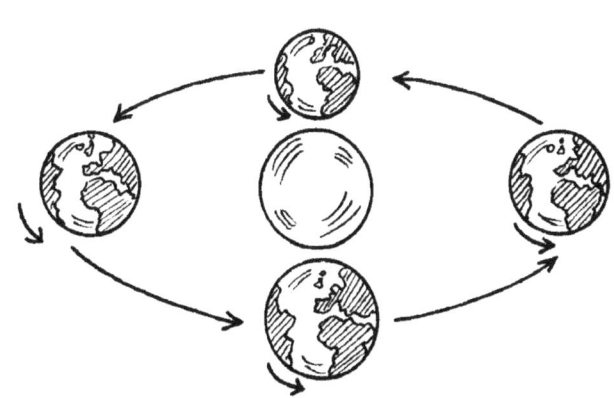

The Earth and Beyond

The twice-daily **tides** which we observe are caused by the Moon's and the Sun's gravitational pull on the water in our seas and oceans. The Moon pulls the water into a slight bulge, causing a high tide on the side nearest the Moon and a low tide on the far side. The highest tides are produced when the Sun and the Moon are pulling in the same direction. When the Sun and Moon pull in different directions, they produce neap tides.

Some ideas for further work

◆ Find out which countries have more than one time zone. What problems do you think could be caused by having more than one time zone in the same country?

◆ Find out about space probes to other planets or their moons. Are there any probes already on their way to other planets or moons and, if so, when should they arrive and what will they do?

◆ Build a 3-D model of the solar system. The Sun and planets could be made by covering balloons with papier-mâché. Inflate the balloons according to scale. Paint them in their natural colours and hang them from the ceiling, separated according to scale.

The Earth and Beyond

Planet	Atmosphere	Time taken to orbit Sun in Earth years	Average distance from the Sun in millions of km	Diameter in thousands of km
Mercury	No atmosphere	0.24	5.8	4.9
Venus	Carbon dioxide	0.61	108	12.1
Earth	Nitrogen, oxygen	1.00	150	12.8
Mars	Carbon dioxide	1.88	228	6.8
Jupiter	Hydrogen, helium	11.86	778	142.8
Saturn	Hydrogen, helium	29.50	1427	120.0
Uranus	Hydrogen, helium	84.00	2875	51.1
Neptune	Hydrogen, helium	164.80	4496	49.5
Pluto	Methane	248.40	5900	4.0

The Earth and Beyond

Activity	Answers	Resources needed	Teaching/ safety notes
Sheet 1 (page 98) The shape of the Earth	• Yuri Gagarin (1934–1968) • Neil Armstrong (Apollo 11, 1969) • The Apollo • 'That's one small step for man, one giant leap for mankind.' (Neil Armstrong)	Reference materials	In Neil Armstrong's words 'a' before 'man' was missing in the live transmission but was inserted in later recordings.
Sheet 2 (page 99) My mother and the pizzas	Moving out from the Sun we should have: Mercury, Venus, Earth, Mars, Asteroids, Jupiter, Saturn, Uranus, Neptune, and Pluto	Reference materials	For many years it was thought that the space between the orbits of Mars and Jupiter was occupied by another planet. It has since been discovered that it is littered with thousands of stony or metallic boulders called asteroids. These sometimes collide with the Earth as meteorites.
Sheet 3 (page 100) The shape of the planets	• The Earth is like a giant spaceship because it is travelling through space with us on it as passengers. Like a spaceship, the Earth is filled with everything we need to survive during our journey. • Jupiter has a giant red spot which is thought to be a storm which has raged for hundreds of years. • Voyager (USA space probe) has discovered that Saturn's rings are narrow bands (500 m) of ice.	Reference materials	As the Earth and the other planets orbit the the Sun, the Sun and the planets orbit the centre of the galaxy. It takes around 225 million years for the Sun to orbit the centre of the galaxy.
Sheet 4 (pages 101, 102 and 103) The size of the planets	Some suggested colours: Mercury pink Venus grey/black swirling clouds Earth blue/green Mars red Jupiter yellow and white stripes Saturn orange and yellow bands Uranus greenish/blue Neptune bluish Pluto yellow	Resource sheet Scissors Stiff card String Sticky tape Cardboard tube	For many years it was thought that Venus could have a similar atmosphere to Earth's (hence the sci-fi films of the 1950s). Probes, however, have shown Venus to be a hostile planet with sulphuric acid rains and (very) high temperatures.

The Earth and Beyond

Activity	Answers	Resources needed	Teaching/ safety notes
Sheet 5 (page 104) The Sun and the Earth	• Midday: Position 2 • Dawn: Position 1 • Sunset: Position 3 • Midnight: Position 4	Globe Plasticine (optional)	Make little Plasticine people and stick them around the equator. Use a torch to represent the Sun and spin the globe anti-clockwise.
Sheet 6 (page 105) Shadow lengths	• The length of the shadow will depend on the height of the stick, the season and the position of the Sun. • The shadow should be longest at the beginning and the end of the day. • The shortest shadow should occur around midday.	Stick Tape measure	Do this activity on a sunny day.
Sheet 7 (page 106) Sundial	The shadow should vary in length and direction throughout the day as the Sun's relative position changes.	Stiff card Blu-tack Scissors	A sundial would normally have a North–South pointing, triangular shape which casts the shadow.
Sheet 8 (page 107) Night and day	Emergency service staff all work at night as well as during the day; night watchmen. Badgers and foxes are also nocturnal animals.	Reference materials	
Sheet 9 (page 108) Times round the world	• Japan: 12 o'clock • USA: 11 o'clock • Finland: 8 o'clock • Brazil: 5 o'clock • Australia: 9 o'clock		Large countries such as the USA have several time zones, so it is necessary to mention a location in order to work out the relevant local time.
Sheet 10 (pages 109 and 110) Phases of the Moon	Over a period of 30 days the children should be able to record the phases of the Moon and realize that they form a repeatable pattern.		The Moon takes 27 days and 8 hours to orbit the Earth. As the Moon orbits the Earth it also spins on its axis. The time taken for the Moon to turn completely around is also 27 days and 8 hours. This means the same side of the Moon always faces the Earth.

Physical Processes

The shape of the Earth

The Earth and Beyond — Sheet 1

 **Read** For many years people were unsure about the shape of the Earth.

There were lots of disagreements!

Look further Use reference books, CD-ROMs or the Internet to answer these questions.

◆ Who was the first person in space?

◆ Who was the first person to walk on the Moon?

◆ Name the spacecraft which took people to the Moon.

◆ What were the first words said from the surface of the Moon?

The Earth and Beyond
Sheet 2

My mother and the pizzas

 On nights when there is a full Moon, we can see it is a sphere. Its shape is just like the Sun, the Earth and all the planets. The Moon orbits the Earth. The Earth orbits the Sun along with the eight other planets.

 Label below the nine planets orbiting the Sun.

Venus	Mars	Pluto
Earth	Saturn	Uranus
Mercury	Neptune	Asteroids

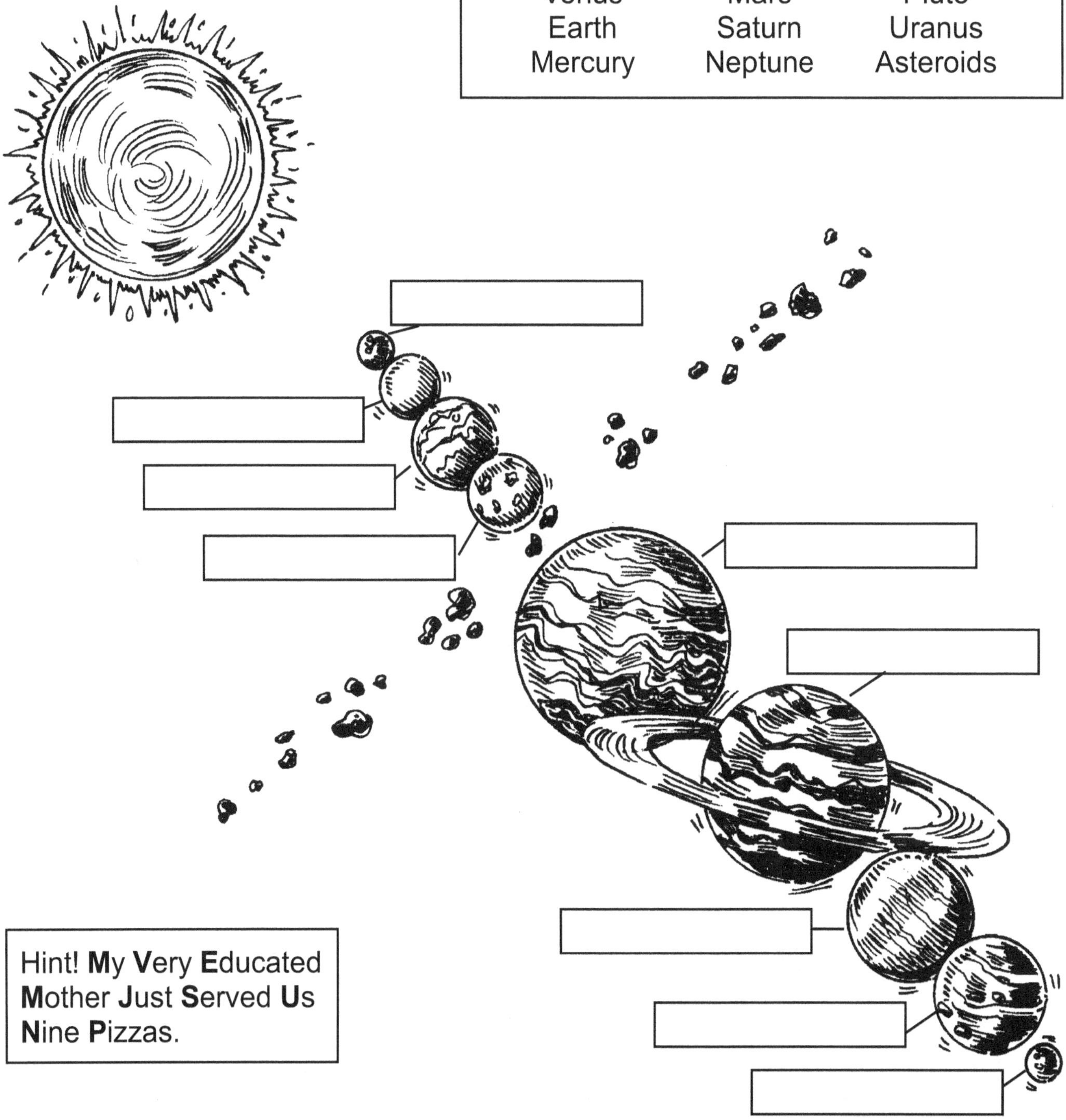

Hint! **M**y **V**ery **E**ducated **M**other **J**ust **S**erved **U**s **N**ine **P**izzas.

© M Abraitis, A Deighan, B Gallagher, B Smith & M Toner

The shape of the planets

The Earth and Beyond — Sheet 3

 **Look further** — Use CD-ROMs and/or reference books to find the answers to these questions.

	Can you think why some people call our planet a giant spaceship?
	This picture is of a planet which has a large red spot. Find out the name of the planet and what causes the spot.
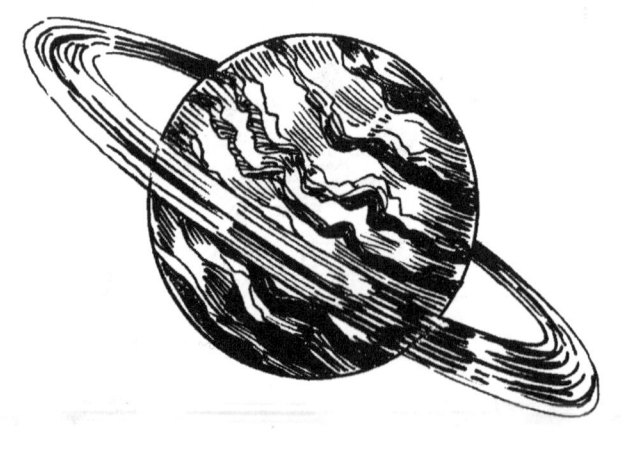	This picture is of a planet which has lots of rings. Find out the name of the planet and what causes these rings.

Physical Processes © M Abraitis, A Deighan, B Gallagher, B Smith & M Toner

The Earth and Beyond — Sheet 4: The size of the planets

You need: Resource sheet (The size of the planets), scissors, stiff card, string, sticky tape and a cardboard tube (eg a kitchen roll tube).

 Follow these instructions to make a scale model of our solar system.

1. Cut sets of holes in the centre of a cardboard tube as shown.

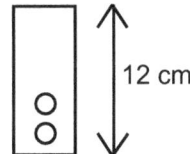

2. Cut two lengths of stiff card and take care to measure and mark them as shown.

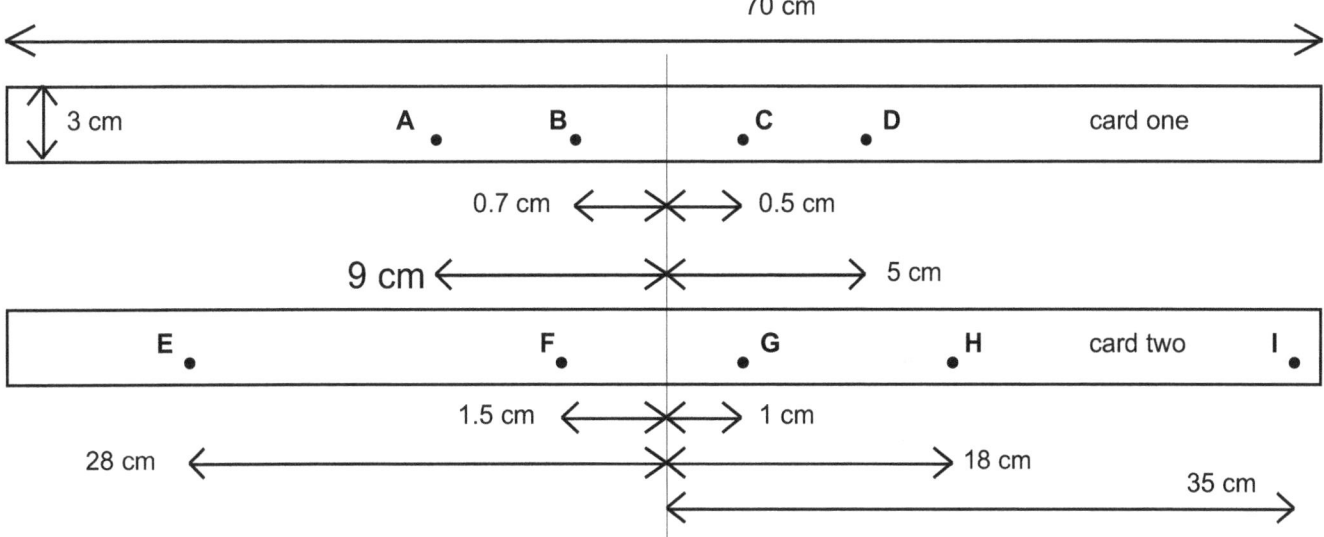

3. Roll each length of card into a long tube and fix them together using sticky tape at each end.

4. Slide tube one through the top hole in the main central tube and tube two into the bottom hole.

5. Cut out the planets from the resource sheet and colour them.

6. Glue the planets and their name tags onto stiff card.

7. Cut nine pieces of string 10 cm long and hang the planet names below the corresponding planets.

The size of the planets

The Earth and Beyond
Sheet 4 – Continued

8   Attach the Sun 10 cm below the central cardboard tube.

9   Use the table below to find out which planet to attach at each point.

Point	Planet
A	Saturn
B	Venus
C	Mercury
D	Jupiter
E	Neptune
F	Mars
G	Earth
H	Uranus
I	Pluto

10 Hang the completed mobile up with string.

Your mobile will give you an idea of the distance of each planet from the Sun and the size of the planets compared with each other.

The size of the planets

The Earth and Beyond
Sheet 4 – Resource sheet

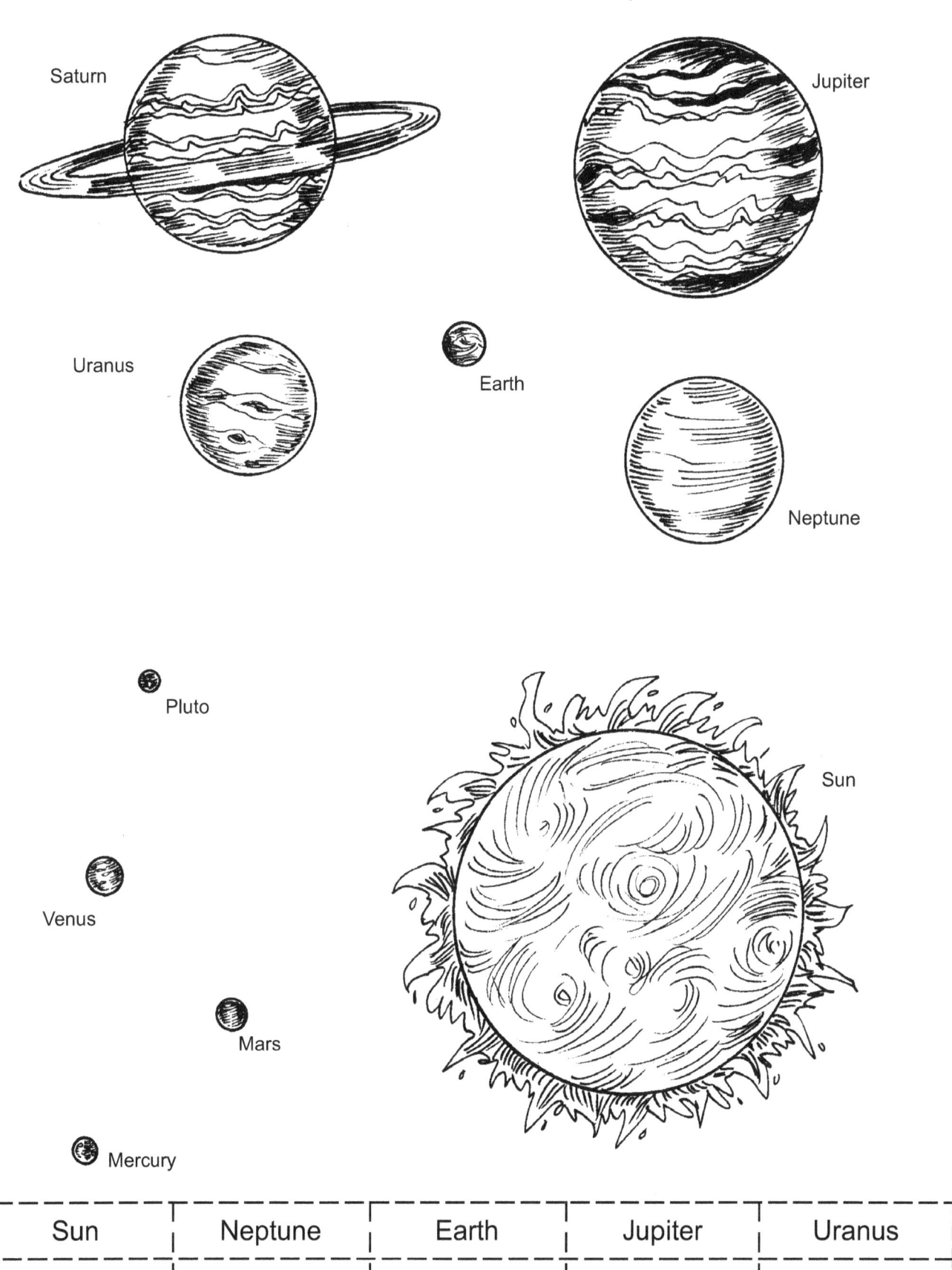

| Sun | Neptune | Earth | Jupiter | Uranus |
| Saturn | Pluto | Mars | Mercury | Venus |

The Sun and the Earth

The Earth and Beyond — Sheet 5

You need: a globe of the Earth.

 **Read** — It takes the Earth one day to complete one rotation. The Sun does not move but, as the Earth spins, the Sun appears to rise and fall during the day.

 **Activity** — Use the globe to show how the Earth continually spins with us on it.

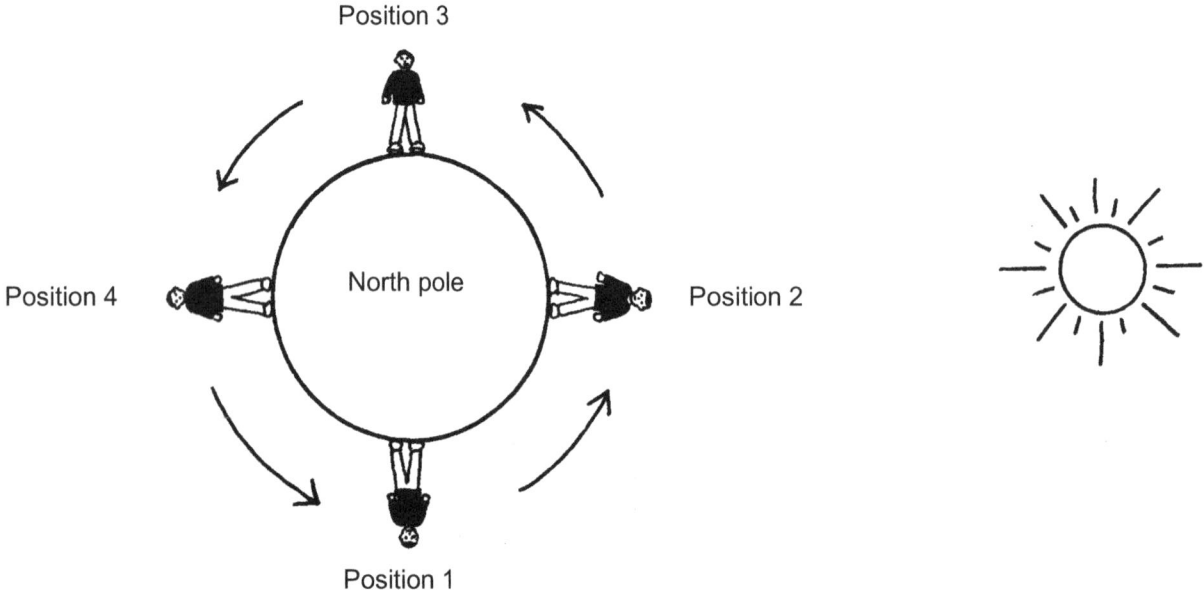

Match the position of the person during the day with the correct time of day.

Time of day	Midday	Dawn	Sunset	Midnight
Position				

The Earth and Beyond
Sheet 6

Shadow lengths

You need: a sunny day, a stick and tape measure.

- ◆ Place a stick in the ground. Measure the length of the shadow caused by the Sun as the Sun changes position during the day.
- ◆ Record your results in the table below.

Time of day				
Length of shadow				

Make a bar chart to show how the length of the shadow changes throughout the day.

The Earth and Beyond
Sheet 7

Sundial

You need: stiff card, a pencil, some Blu-tack and scissors.

Read

As the position of the Sun changes throughout the day, the shadows it casts change in size, shape and position. Before modern clocks, people used to tell the time by looking at the position of shadows.

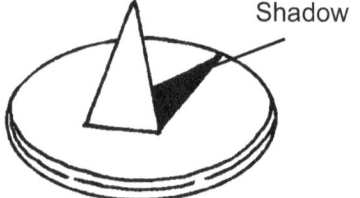
Shadow

Activity

Let's build a sundial.

◆ Cut out a circle of 20 cm diameter from the stiff card.

◆ Stick a pencil in the middle of the card with the Blu-tack.

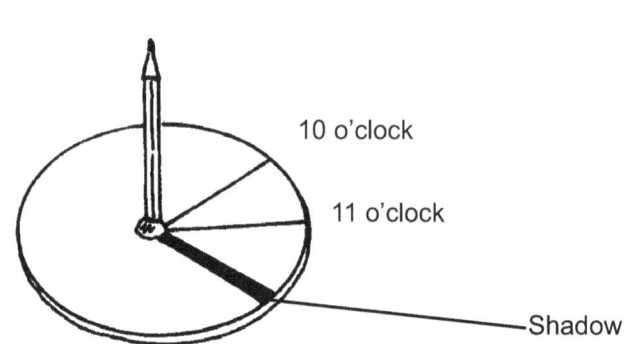
10 o'clock
11 o'clock
Shadow

◆ Put the sundial in a sunny place, either outside or at a window through which the Sun shines.

◆ Mark the position of the pencil shadow every hour. You may need to do some measuring at breaktime or lunchtime, but this will make the science more accurate.

◆ Use your sundial to tell the time the next day it's sunny. Ask your teacher to use your timing for the different lessons of the day.

Look further

Can you find a sundial near where you live? You could try looking in the park or on the side of a church tower.

Physical Processes

© M Abraitis, A Deighan, B Gallagher, B Smith & M Toner
This page may be photocopied for use by the purchasing institution only.

The Earth and Beyond
Sheet 8

Night and day

 **Read** We live on planet Earth. It spins round completely once every 24 hours. We say that the Earth rotates once every 24 hours. Daytime is when our part of the Earth faces the Sun. Night-time is when our part of the Earth does not face the Sun.

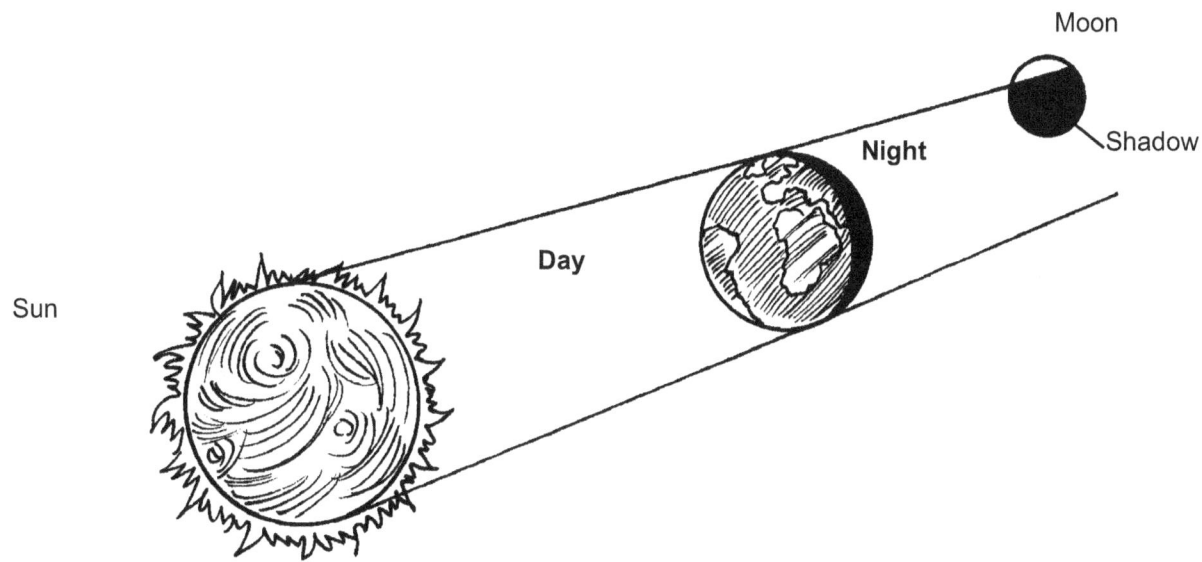

 **Activity** Think of examples of people and animals who work, feed or hunt during the day or at night. The first ones are done for you.

	Daytime	Night-time
People who might work during the day or at night	A teacher	A policeman
Animals which come out to feed or hunt during the day or at night	A rabbit	An owl

The Earth and Beyond
Sheet 9 — Times round the world

 As the Earth rotates, different parts of it receive light from the Sun. This makes the time different. For example, it can be midday where you live, but midnight somewhere else!

This table shows how the times in other countries around the world are different from the time in the United Kingdom.

Country	Flag	Time difference from UK
USA (east coast)		5 hours behind
Japan (east coast)		10 hours ahead
Australia (central)		9 hours ahead
Finland		2 hours ahead
Brazil (east coast)		3 hours behind

 For each clock, work out the time in each of the countries.

Time in UK	2:15	3:00	6:00	4:30	12:00
Find the time in:	Japan	USA	Finland	Brazil	Australia

The Earth and Beyond
Sheet 10 — Phases of the Moon

 **Read** We see the Moon because it reflects light from the Sun. As the Moon orbits the Earth, the amount of light it reflects also changes. These are called the phases of the Moon.

 Full moon		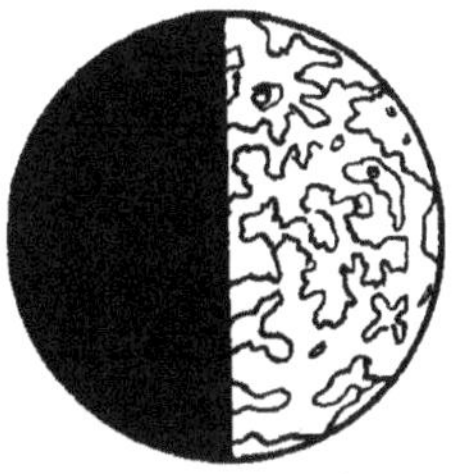 Half moon	
 Crescent moon			 Full moon

 **Activity** Use the chart on the next page to draw the phases of the Moon over the next 30 days.

Phases of the Moon

The Earth and Beyond
Sheet 10 – Continued

Answer

For the next 30 days draw in the shape of the Moon each night. (On cloudy nights you may not be able to see the Moon.)

Night 1	Night 2	Night 3	Night 4	Night 5
Night 6	Night 7	Night 8	Night 9	Night 10
Night 11	Night 12	Night 13	Night 14	Night 15
Night 16	Night 17	Night 18	Night 19	Night 20
Night 21	Night 22	Night 23	Night 24	Night 25
Night 26	Night 27	Night 28	Night 29	Night 30

Glossary

These definitions, while accurate, are by no means comprehensive. If you require a more exact definition, you should consult a good scientific dictionary.

air resistance
The force which pushes/pulls against a moving object as it moves through the air.

attraction
A word used to describe a force which pulls objects together, for example the magnetic force between a paper clip and a magnet.

battery
A portable source of electricity, a battery contains chemical energy which is converted to electrical energy when it is used in a complete circuit.

circuit
A word used to describe the way in which a source of electrical energy and a component(s) is connected together by wire conductors.

conductor
In electrical terms conductors are materials such as metals which allow electricity to flow through them easily, without the generation of too much heat.

electricity
A form of energy which is used to power many devices such as computers and household appliances like washing machines, televisions, lights and fridges.

energy
Energy allows us or machines to do work. For example, we receive our energy from the food we eat which is chemical energy. We are then able to change this energy to do work.

force
An influence which can change an object's shape, direction or speed.

forcemeter
A device for measuring the size of a force.

friction
The force which pushes or pulls against a moving object's motion.

gravity
The force of attraction between objects with mass. For example, the Earth's gravity attracts other masses down towards its centre.

insulator
In electrical terms an insulator is a material that does not allow electricity to pass through it easily; this is called resistance.

loudness
The amplitude or height of a sound wave.

moon
A natural mass which orbits a planet.

opaque
A word used to describe materials which do not allow light to pass through them.

orbit
The circular or elliptical motion of planets around stars, and moons around planets.

parallel circuit
An electrical circuit with two or more components such as bulbs connected individually to a source of electricity. Each component can be switched on or off independently of each other.

phases
The description of the Earth's shadow on the Moon during the Moon's monthly orbit.

pitch
The tone of a sound wave. A high-pitched sound has a high tone. This is directly related to the number of vibrations per second produced by the source which is creating the sound.

planet
A natural body which orbits a star such as our Sun.

poles
The two ends of a magnet at which the attraction appears to be the strongest. The Earth could be described as a magnet, with the strongest attractions at the North and South poles.

reflection
In terms of light, reflection occurs when light bounces off a material. We see things either because they give off their own light, such as fires, or more commonly because light reflects off them.

repulsion
A word used to describe a force which is pushing objects apart. For example, two like poles of a magnet will repel each other.

rotation
The spin of an object such as a planet around its axis. The Earth takes 24 hours to complete one spin on its axis.

series circuit
As opposed to a parallel circuit, in this type of circuit the electrical components are connected in a line to the same source. If one component is switched off or broken then this will affect the operation of the other components.

shadow
The area behind an object on which little or no light has fallen.

solar cell
A device by which electricity can be generated directly from sunlight or any other light source. They are commonly used to power calculators, etc.

source
The origin of a form of energy such as sound, for example a musical instrument.

space
The vacuum between particles of matter and astrological bodies such as planets and stars is occupied by space.

streamlining
A name given to the shaping of objects which is intended to reduce the force of friction acting against them when they are moving.

Sun
The Sun is the name given to the star at the centre of our solar system.

sundial
A device which is used to judge the time of day by measuring the position of an object's shadow.

translucent
A material which allows light to pass through it, but does not allow a clear image to be seen.

transparent
A material which allows light to pass through it, but also allows a clear image to be seen.

upthrust
Any force which pushes upwards. It can be noticed in boating where the upthrust from the water counteracts the force of gravity.

vibration
In terms of sound this describes the regular motion of a moving object, such as a string on a guitar, which is producing the sound.

www.ingramcontent.com/pod-product-compliance
Lightning Source LLC
Chambersburg PA
CBHW081350160426
43197CB00015B/2720